Cultivating Community

Interest, Identity, and Ambiguity in an Indian Social Mobilization

Michael Youngblood

Cultivating Community

Interest, Identity, and Ambiguity in an Indian Social Mobilization

Michael Youngblood

South Asian Studies Association
Pasadena, California
www.sasaonline.net

Published by SASA Books
A Project of the South Asian Studies Association
Pasadena, California 91367
A public benefit, non-profit corporation, EID 26-1437834
www.SASAonline.net

ISBN 978-0-9834472-7-6 (paperback)
LCCN: 2016950517

SASA Books is a project of the South Asian Studies Association, a recognized 501(c)3 non-profit, public benefit corporation of scholars and others interested in South Asia.

Using a pro bono model, SASA Books is dedicated to publishing high quality scholarly materials using a rigorous double blind vetting process. The SASA website is www.SASAonline.net.

Contents

Maps and photographs

Preface

This book has been a long time coming. It was a labor of love written over the course of several years, but then I moved on to other endeavors and never published it. Perhaps this is a fitting journey for a book that is largely about shifts and paradoxes in human behavior.

A number of nudges have led me to publish this book now. One is that it has continued to be a touchstone for me over the past ten years. As an anthropologist working in human-centered design an important part of my charge is to understand ways that people and communities engage their material worlds, ideas, and each other. My years of fieldwork with the Shetkari Sanghatana are a reservoir of experience that has helped me in this work. Whether looking for opportunities to improve communication between public health agencies and underserved populations, working with community stakeholders to create a shared vision for the future of their public schools, or designing programs for teen leadership development, I have often found myself drawing on insights first gained during my fieldwork in small towns and villages across Maharashtra.

I have also been nudged by India's changing position in the world. In the years since my fieldwork with the Shetkari Sanghatana, India has surpassed the one billion population mark, has emerged as a center for high-tech industries, and is becoming an increasingly important global player. Maharashtra state, which is about the size of Poland but has three times the number of people, is one of the leaders in the new Indian global economy. But, while the rise of industrial India is fascinating, this book is focused on another side of the Indian story—one far removed from urban India but not disconnected from it. Despite pockets of surging wealth and global influence in India's cities, today most Indians live in rural areas and earn their livelihood from the land.

While some rural Indians are relatively well off, the majority toil in small fields using bullocks or other beasts of burden to pull plows that have not changed markedly in hundreds of years.

The human numbers alone are a good argument for paying regard to the rural Indian experience. Roughly seven out of every ten Indians live in a village and depend directly or indirectly on agriculture for their livelihood. To put this in perspective, that translates into nearly one in nine human beings on the planet. It behooves us, as social scientists and as global citizens, to further our understanding of such a prominent share of the human experience.

A third nudge has been the surge in public attention to mass social mobilizations in the United States and around the world—and the apparent simplicity with which governments and media have framed their attitudes and responses to these. The narratives through which influential opinion-makers in the Western world have represented mass phenomena as diverse as, for example, the Arab Spring movements, the Occupy movement, global movements in response to climate change, or even ISIS and Al Qaeda have struck me in so many ways as surface-level, generalized, and incomplete. That people may hold these simplistic narratives in mind while making local and international policy decisions is even more unsettling. Of course, these phenomena are all different from each other and from the Shetkari Sanghatana. They voice distinct philosophies and goals, and exhibit distinct ways of acting in the world. I am not an expert on these other movements, but our ability to think about them seems hamstrung by a focus on their loudest voices, their public pronouncements, and their most attention-getting acts. The Shetkari Sanghatana taught me something that I believe to be a general truth: mass social movements, no matter how monolithic or unified they may seem, contain within them substantial ideological ambiguity and wide-ranging motivations for participation. Sociologically, they are far more complex than how their leaders and propagandists represent them to the outside world.

Of all the nudges, however, there is one that has most compelled me to finally publish this book: people have asked for it. Thanks to a few accolades that were bestowed on an early manuscript and the PhD dissertation that preceded it, people have from time to time contacted me to find out where they can get the book. Until now, the answer was "nowhere." I thank these inquisitors for pushing me.

A note on facts and figures

The Shetkari Sanghatana continues to be active in Maharashtra, though times have inevitably changed and the movement has evolved. Trying to tell the most current story of a mass mobilization in a printed book sets up a moving target, the story is always passing into history by the time a book is published. This book is a field study, conducted at a particular period in time— between 1996 and 1999. I have not attempted to significantly update my account of the movement since completing the first manuscript nearly ten years ago. Facts and figures, unless indicated otherwise, are those that were available at the time I wrote the original manuscript. Citations of publications and authors reflect the literature of that same time and not the most current theories or debates in relevant fields. Readers hoping to glean the very latest in Indian field data or social theory will be disappointed. My hope is that readers will approach this work as a product of its time and a rich examination of complex human phenomena that lends insight to our world, both yesterday and today.

Protecting informants

Ethnographic ethics and conventions require the protection of informants. In the ensuing pages, I have sought to preserve the anonymity of informants in the following ways.

Regarding names of people and the villages in which they live, I have chosen to always disguise these through the use of pseudonyms or vague references. For example, I may refer to an informant as a "senior organizer in Marathwada region" or to a location as "a village in Jalna district"—but will rarely be more specific.

Throughout these pages there is only one informant whom I will normally refer to without disguising his name: Sharad Joshi. As the recognized paramount leader of the movement and its chief spokesperson at the time of my field research, Sharad Joshi's opinions and comments are, for the most part, a matter of public record. Thus, wherever I have attributed comments to Joshi, I have treated them in this manner. To be sure, many of my discussions with Joshi were of a more candid and "off the record" nature. Where this was the case, I have maintained his privacy.

Translation, orthography, and use of non-English words

Sharing linguistic detail across a diverse readership is a challenge. For the sake of readability, I have tried to follow conventions that provide sufficient detail for the specialist while still being accessible to a general audience. Wherever I have used Marathi, Hindi, Sanskrit, and other Indian language words I have transliterated these into standard Roman script according to the following guidelines:

- Proper names, such as people, deities, and place names that have no English alternatives and would be used in normal English speech without translation are treated as English words. For example: "Gandhi," "Vishnu," or "Mumbai."
- Broad, generalized terms for social communities and varnas are also treated as English words, but generally not as proper nouns. For example: "brahmin," "shudra," (as opposed to Brahmin, Shudra). Comparatively specific caste (jati) names, however, are treated as proper nouns. For example "Maratha" and "Mahar" (as opposed to maratha or mahar).
- In general, I have followed English rules for pluralization, capitalization, and possessive constructions for all non-English words that lie within English sentences. The exception to this is words and sentences

that have been transliterated as non-English text or quotations. These have been left grammatically intact.

- Indian language orthography often allows for variable renderings of words into Roman script. Where this is the case, I have either used the most common Anglicized spelling or the spelling that most closely approximates the spoken pronunciation with which I am familiar. Thus for example the Devanagiri character "व" might appear as a *w* or a *v* when written in standard Roman script. I have used both, depending on what I understand to be the common Anglicized spelling or the most common pronunciation. For example, I have chosen *Varkari* as opposed to *Warkari*, but also *swadeshi* as opposed to *svadeshi*.

- For simplicity, in the majority of cases, I have omitted the silent *a* that follows most Marathi words and names ending in a consonant.

All translations from original Marathi speech or writing are my own, unless indicated in the text or a note.

Acknowledgments

This book could not exist without the contributions of many people and institutions. I thank my editor, Susan Walters Schmid, for helping me prepare the final manuscript for publication, and my wife Mary Ellen for putting up with me throughout the process. But these are merely the most recent contributions. While doing field research and writing the PhD dissertation that preceded this book, I was fortunate to have input and insight from Kirin Narayan, Katherine Bowie, and Anatoly Khazanov, all in the Department of Anthropology at the University of Wisconsin–Madison. Kirin Narayan was a particularly important source of encouragement, and I am very appreciative of the many hours that she dedicated to reading and providing feedback.

During my research in Maharashtra, I was affiliated with the Department of Politics and Public Administration at the University of Pune. I am grateful to the faculty and staff of that department, especially my late friend and mentor Rajendra Vora, for his expert advice and for facilitating introductions to other scholars, students, and individuals associated with the Shetkari Sanghatana.

In the course of my research I also received invaluable assistance, support, insightful critiques, suggestions, and companionship from many other institutions and individuals. Sadly, there have been instances in the not-so-distant past in which Maharashtrians have been harassed for their friendship or intellectual engagement with foreign scholars. To protect the welfare of all, I am disguising the names not only of my informants (which is standard anthropological practice) but also of the many friends, mentors, surrogate family members, field assistants, language tutors, translation assistants, and others that I would prefer to acknowledge by name. In particular, among them I must acknowledge G. O., G. P., Kumar S., M. K., V. B., and Vinay H. for their time, insight into the Sanghatana and other movements, and hospitality on so many occasions. For help with transcriptions and understanding the intricacies of Marathi folklore and culture, I am indebted to Arati K., D. C., H. R., M. J., N. B., S. B., S. P., S. R., V. D., and V. K. For general hospitality, support, and insight, my particular thanks to Mrs. A and family, G. P. and family, Kumar S. and family, Srikant P. and family, Sudhir W. and family, L. S., P. K., and J. L. There are of course many other people without whose time and companionship my experience would not have been nearly as productive or enjoyable. If I have failed to thank you here, please know that I am extremely appreciative.

Out of all the acknowledgements I have to make, very important among them is the Shetkari Sanghatana itself. Though I never projected myself to Sanghatana leaders and participants as anything but an impartial observer, this organization welcomed me warmly and allowed virtually unrestrained access to its people, activities, meetings, and archives. It not only gave me the space but, remarkably, also the *encouragement* to view the organization and the movement with a critical eye. For this, I must ultimately thank the late Mr. Sharad Joshi, the then paramount leader of the movement, who enabled this with his explicit and implicit consent. I also extend deep appreciation to Mr. Sureshchandra Mhatre, with whom I spent many days as a guest at Angarmala—the movement's village headquarters—and engaged in very many hours of stimulating conversation over the course of my fieldwork. In addition, I must give my thanks to all the men and women of the Shetkari Sanghatana, who welcomed me wholeheartedly as an observer and who shared their lives with me. Most of them will never see this work or even know the extent to which their thoughts and experiences have been essential to the final product.

Finally, I am grateful for fieldwork support that I received from a Fulbright-Hays doctoral fellowship and an American

Institute of Indian Studies dissertation fellowship. Without that support, this project would never have become reality.

Credits

Cover: Cover design by Kathleen Cunningham. Cover photo by the author.

Maps and photographs: All maps and photographs by the author, with the following exceptions: Photo 2 ("Participants raise their fists") on page 12 courtesy Gunvant Patil Hangargekar; Photo 14 ("Vishnu saving the world from Bali") and Photo 15 (Detail) on pages 222–23 courtesy Cathleen Cummings.

Portions of chapter 5 were published previously by Oxford University Press as "Negotiating Hierarchy and Identity: Cultural Performances on the Meaning of the Demon King Bali in Rural Maharashtra" in *Speaking Truth to Power: Religion, Caste, and the Subaltern Question in India*, edited by Manu Bhagavan and Anne Feldhaus (2008).

1—Introduction: Interests and identity in a mass movement

Zindabad! [Victory![1]]
Shetkari Sanghatana Zindabad!
Sharad Joshi Zindabad! [Long live Sharad Joshi!]

—Cheer from an audience of 30,000 attendees at a
Shetkari Sanghatana rally,
December 1996, Akola, Maharashtra

On a chilly winter morning in 1997, an organized group of peasants and farmers in Maharashtra state, India, armed themselves with monkey wrenches and wire cutters. Addressing each other collectively as *shetkaris*[2]—the caste- and class-ambiguous term for "agriculturalist" in the local Marathi language—they gathered in a field marked by tall electrical towers and high tension power lines in the state's eastern, less prosperous, and overwhelmingly rural region called Vidarbha. For seventeen years, the mass movement in which they participated had been experimenting with a variety of tactics to pressure the state government into meeting their demands. Under the banner of their movement, called the Shetkari Sanghatana (the agriculturalists' union[3]), they had closed off their villages, denying entry to outsiders. On other occasions they had refused to send their produce to market. They had thrown onions at politicians and bank officials, or paraded them around for public humiliation on the backs of donkeys. And they had stopped traffic on crucial railroad and highway routes for days on end throughout the state. In this, their latest and perhaps most dramatic bid to bring the urban-based government they called the "black British" to the negotiating table, the Shetkari Sanghatana intended to disconnect one of the state's most critical conduits for electric power—the high voltage lines that run from Vidarbha far across the state to fuel the economy of India's and Maharashtra's most important industrial city, metropolitan Mumbai.[4] The Sanghatana gave an advance warning, printed for the public to see in the *Indian Express,* a newspaper popular with urban, English-literate Maharashtrians. In it the black British were given a choice: either negotiate with the shetkaris or pay the consequences (*IE* 12/4/96). When the government failed to respond, the Sanghatana prepared to make good on its threat.

Early on the morning of January 6, the activists assembled and made a solemn declaration. Henceforth, they announced, the electric towers were no longer part of India. Like so many of the villages that surrounded them, the towers now belonged to another, unrecognized country lying within India's borders—a country once ruled by an honorable, god-like demon named *Bali,*[5] whose kingdom of villages and shetkaris is known as *Bharat.*[6] After a brief ceremony, waving banners bearing Marathi slogans of "cut power to Mumbai," and "Bali will return," the volunteers began scaling the high-voltage towers. Untrained for the job, many of them risked electrocution. All risked confrontation with the police who would appear on the scene to arrest them and beat them with batons. As the morning wore on, at least one activist met his death while attempting to disconnect the power lines.

This was not the first time that the nonviolent and unarmed participants in the Shetkari Sanghatana had put their lives and livelihoods at risk in the name of Bali's kingdom. Nor was it the first time the Sanghatana had lost comrades in the struggle. In one of its first major road blockages in 1980, two shetkaris were shot and killed by the police, hundreds were wounded by blows from police batons, and thousands were sent to prison (Sahasrabudhey 1989).[7] In two other actions the following year, sixteen shetkaris had died in confrontations with the police, and hundreds of others were beaten or taken to jail (ibid.). These risks are well known to the shetkaris. Those who have died in the struggle live on as martyrs in conversation, posters, songs, and poems. Almost two decades later, gathered around the electricity towers, activists were still risking all in the name of Bali and of the movement's charismatic leader—Sharad Joshi—a man whom many of them profess to be the incarnation of Bali himself.

A few days before the attempted unplugging of Mumbai, I was drinking afternoon tea in the shade of a banyan tree, talking with a Shetkari Sanghatana activist named Sandeep.[8] He and other local villagers had been speculating on the impact the electricity agitation might have. "I'm not sure if this will move the government to listen," he said, "but our men and women are willing to die trying." When I asked him why, he began with a familiar explanation I had heard many times during my two years of research on the Shetkari Sanghatana. He explained how the government systematically expropriated wealth from the rural areas of Bharat in order to fuel the economy of the cities. He described how the economic agenda of the movement—known as the One Point Plan—calling for higher prices on agricultural produce addressed the challenges faced by everyone who lives in the villages of Bharat. He also described the leadership and charisma of Sharad Joshi, the movement's chief strategist and ideologue. "Sharad Joshi," he said, "is a god for us."

Sandeep looked at me for a moment as if concerned that I had not weighed the full import of his words. Putting down his teacup, he grabbed my hand and led me down a narrow lane. We took a short walk through the village and entered a home. There he directed me toward the family god house—a small, rectangular shelf recessed into a chipped, oil smudged, blue-painted concrete wall. By the light of a dusty bulb hanging dim over the room, Sandeep pointed to a photograph of Sharad Joshi nestled among the symbols and figurines of the gods.

What is going on here? We see a deified human leader, a millenarian call to restore a glorious and mythic past, mobilized rural populations willing to die in a struggle against elected government and modern urban life, and soaring narratives of martyrdom about those who have fallen. These snapshots of social rebellion summon up apparent analogues in many parts of the developing world. But to understand the Shetkari Sanghatana merely through such snapshots is to misunderstand a great deal about how social movements emerge and proliferate, how they are perceived and experienced by their adherents, and how they are entwined with broader social processes in the world around them.

This book explores the motivations and subjectivity of participants in a mass social movement. It is an exploration in human agency and the intricate webs of factors within which people make choices and build agentive relationships that affect their lives and their societies in myriad and important ways. More pointedly, it is a response to suggestions—implicit and explicit in much of the popular and scholarly writing on social movements—that the political behaviors and allegiances of agrarian or "non-Western" subjects, far more so than those of their urban, Western counterparts, tend toward ideological and non-rational extremes. The Shetkari Sanghatana offers a case in point.

Let us examine the claim more closely. When we look at the sociological literature on social movements we see that much of it approaches these movements as phenomena characterized by ideological consensus and a loyal body of participants. This is what I call the *fallacy of solidarity*. My contrary argument is that successful mass social organizing may often be better understood as characterized by contextually shifting participant agendas and fluctuating memberships. In other words, quite the opposite of being rigidly defined and ideologically hegemonic, large-scale social movements may be relatively ambiguous. They may draw on a pool of multivocal symbols; they may be subject to wide variances of interpretation by their participants; and they may be socially and geographically amorphous. This ambiguousness is not necessarily a failing or a hindrance for mass movements. As I will argue, ambiguity can actually enable their expansion and sustainability, as it does for the Shetkari Sanghatana. Moreover, ambiguity does not necessarily signal a lack of thoughtfulness or political consciousness among individual participating agents.

On the contrary, Shetkari Sanghatana participants are highly aware of the ambiguity in the key ideas of the movement's overarching ideology, as well as the ambiguity of their own "membership" and action in the movement. To a large extent, as I shall argue, participants are actually empowered by the interstices of meaning that ambiguity affords.

These conditions of large, organized social movements such as the Shetkari Sanghatana were not adequately accounted for by most social theory at the time of my research. Overwhelmingly, attempts to identify the salient aspects of "successful" movements have portrayed their agendas and, particularly, their ideology as clearly defined and broadly embraced among participants—or, implicitly at least, as evolving toward that end. There is in this a subtle and generally unstated premise that we could call *progressive solidification*. What I mean by this is the a priori assumption that successful social mobilization corresponds with maximized—or solidified—shared identity and shared felt-interest among the movement's participants. Of course, some degree of convergent identity and interest is a necessary foundation for an organized movement. The premise of progressive solidification, however, takes this too far and leads us to other assumptions that do not stand up to scrutiny.

The premise places us in three perceptual and analytical traps. In the first trap, we are led to expect that strong convergences of interest and identity become increasingly important for holding a movement together as it expands in size and impact. This expectation prevents observers from seeing the numerous, recurring, and far more ordinary instances of nonconvergence that exist at different moments, and in different degrees, across the lifetime (and life-*place*) of a movement.

In the second trap, the premise influences observers to privilege the agenda and ideologies of a movement's most dominant and empowered public voices. These voices are usually those of its central leaders or spokespeople who have the resources and authority to clearly define, in uncomplicated and media-friendly terms, the people and interests that the movement purportedly represents. If we inquire beyond the bounds of the premise, we may find that the true range of participants' interests and identities bears only a partial resemblance to the version espoused by these most empowered voices.

This brings us to the third and most bothersome trap. By dulling our determination to look beyond monolithic representations, the premise of progressive solidarity can lead us to underestimate the rationality and political subjectivity of a movement's actual participants—especially those who are the least empowered, whose behaviors, perceptions, and motivations may seem to be at odds with the view of the world and the vision of change expressed in the uncomplicated, monolithic version of the movement. Despite the desire of social scientists to understand—and even sympathize with—the subaltern experience, the premise of progressive solidification continues to inform conclusions that are partially dismissive of the less empowered. In an effort to explain movements characterized by a broad, diverse social base, theorists may debate whether a movement's ideology is *truly* consonant with the needs of its participants or if, on the other hand, participants have been duped into embracing ideas that serve interests other than their own. In either case, the fallacy of solidarity is largely untouched.

Some social movements seem to offer easy cases in point for a conclusion in which mass participants are hoodwinked by charismatic leadership and hegemonic ideologies. The Shetkari Sanghatana is one of those. This is not simply because so many of its adherents speak of the political landscape in terms of a mythological demon and a semi-deified leader, but also because, for so many of these participants, the central cause for which they appear to struggle and risk their livelihoods seems to be deeply disconnected from their own material interests.

Since it first began making headlines in the early 1980s as India's best organized and most influential agrarian movement, the Sanghatana's leadership has repeatedly and continuously articulated a free-market agenda focused on the deregulation of government-controlled pricing on agricultural produce.[9] In the pronouncements of movement leaders, in the pages and banners of movement propaganda, and even in the public words of most participants, insufficient prices appear to be the single most important issue affecting all agriculturalists. To outsiders, this agenda seems tailored to represent primarily the interests of the state's richest, cash cropping farmers—the ones who grow food primarily to *sell* rather than to feed their families. One cannot deny the apparent paradox that this presents when much of the mass support that has enabled Sanghatana successes has derived from the state's majority of small holding and marginal agriculturalists. These are the precariously positioned and only partially market-oriented cultivators whose larger interests would seem to lie in issues that have been historically more conventional in Indian agrarian politics—such as access to land, protection from local power brokers, debt relief, affordable prices for seed, access to water and fertilizer, and price stabilization to reduce unpredictability and risk. The situation appears even more paradoxical when we consider some of the many other participants in the movement. These include subsistence cultivators, who send little, if any, produce to market, and agricultural laborers, who work for daily wages and have little, if any, productive land of their own. Unexpectedly, the movement even includes rural merchants, who are *buyers* of primary produce rather than sellers. None of these participants would seem to have a primary interest in the capitalist farmers' demand for higher prices for their crops.

Thinking within the constraints of progressive solidification, this social formation of poor peasants and rich farmers agitating together for a strictly market-oriented agrarian future seems to signal a deeply flawed political consciousness and misdirected political agency among many of the movement's participants. If we accept the Sanghatana's most loudly articulated core agenda as the key motivating frame for action, we could easily conclude that participation by any but Maharashtra's wealthiest farmers is an irrational act of systemic reproduction—ushering in, under a pro-market banner, the same systems of unbalanced accumulation and economic hierarchy against which less prosperous Maharashtrian villagers should probably more beneficially fight. And yet, at the time that I began my fieldwork, the Shetkari Sanghatana had been going strong and attracting participants from throughout rural society, for twenty years.

The Sanghatana in brief

The Shetkari Sanghatana movement burst on the scene somewhat spontaneously in February 1980 with an agitation by onion producers in the rich onion region of west-central Maharashtra.[10] Hundreds of onion growers gathered on the road, blocking traffic with their bullock carts and their bodies, refusing to let their onion crop be taken to market or even to let vehicles pass in and out of their villages. These shetkaris had an immediate underlying complaint for their action: they felt that the going rate for their onions was artificially deflated by government price controls. They demanded what they called a remunerative price.

For years, the shetkaris had been obliged to sell their onions, like most other crops of commercial significance in India and Maharashtra, in accordance with government-determined prices. In theory, these prices, set by the national Agricultural Pricing Commission[11] and the Maharashtra State Monopoly Procurement Scheme, are supposed to represent support prices—floor levels below which the produce cannot be bought or sold, guaranteeing a fair return to the producer. In practice, many of the onion growers felt that these prices were effectively lower than what they could command in an open and unregulated market. In response, the shetkaris gathered together and peacefully blocked the road that led from their villages to the market center. They demanded that the government declare a higher procurement price, more in line with what the free market would afford. They also demanded that, ultimately, the government dismantle its intervention in pricing altogether. As leverage, the shetkaris refused to send any of their onions to market, depriving urban consumers of an indispensable daily staple, until the government met their demands. Government, however, did not yield. Unsuccessful, the villagers eventually disbanded from the roads.

This first campaign, though failing to meet its immediate goal, did have the effect of igniting a new political enthusiasm in the region. It also lifted several leaders to prominence—most important among them Sharad Joshi. To most observers, Joshi seemed an unlikely leader of the agrarian masses. Joshi was from the city. He was a brahmin, a bureaucrat, and a former United Nations official who had been stationed in Europe for several years. He often wore blue jeans and golf shirts and sometimes carried a bar of soap in his pocket to freshen up while attending meetings and rallies in villages. Appearing every bit the urban sophisticate, with no personal or family ties to agriculture, Joshi had only recently retired and relocated to the countryside north of the city of Pune and taken up cultivation on twenty-five acres of unirrigated land. Despite apparent odds, Joshi's leadership took root.

In November of the same year, the onion growers, led by Joshi, launched a second effort covering a much more expansive territory across two districts in the state. It was a solid strategic move. Not only did it draw much greater numbers of onion growers into the agitation, but the expansion also encompassed the two areas—Pune and Nasik districts—that constituted the main onion-producing region of the state and together accounted for about 35 percent of total onion production in the country (Sahasrabudhey 1989). Expanding their participation even further, the growers allied themselves with producers of sugarcane, the second major crop in the two districts, who also participated in the agitation. Cane growers tend to be relatively wealthy compared to onion growers because it is a water-intensive crop that can only be grown on well-irrigated land.

This time, about 200,000 shetkaris—men and women cutting across caste, religious, and income groups—participated in a rail and road blockade lasting over a month. Commercial and passenger traffic on major byways was at a standstill; the state began to lose substantial revenue on interstate transport and trade of important crops; and urban Maharashtrian consumers began to complain loudly about the unavailability of onions. The agitation met strong resistance by the state—but despite thousands of arrests, baton charges, and police shootings that resulted in the deaths of four participants and the wounding of many more, the shetkaris refused to disband (Sahasrabudhey 1989). When most of the men were arrested and taken away, women and children gathered on the road or rails to take their place. In the end, the government agreed to a rise in the prices for both onion and sugarcane, and Sharad Joshi was thrust upon the scene as an up-and-coming leader of the agrarian cause.

After this first real success, the movement expanded rapidly. The Shetkari Sanghatana, as it was now well known, staged similar agitations for prices on rice, cotton, tobacco leaf, peanut, chili, gram, and milk as well as further agitations on onion prices. The movement expanded into other regions of the state, eventually taking a firm hold in the eastern regions of Marathwada and Vidarbha, where cotton growers began to clash with police over cotton prices. By 1982, when the Sanghatana gathered for its first formal conference in a small town in Nashik district, approximately 30,000 delegates, representing half the districts of the state, attended the meeting. It was followed by a massive concluding rally with an estimated attendance of as many as 300,000 shetkaris.[12] Sharad Joshi presided over the conference. In his address to the assembled masses, Joshi asserted that the movement, now large, should avoid becoming rigidly structured. It should remain, he argued, as nimble and responsive as possible.[13]

By the late 1980s, the Sanghatana was building effective bridges with rural movements in other states—eventually forming an umbrella organization for like-minded nascent movements, called the Interstate Coordinating Committee (ICC).[14] Though the new price-oriented movements in Maharashtra, Karnataka, Gujarat, and several other states continued to be organized on a regional and linguistic basis, each constructing the names, ideologies, and symbolism of their movements in local and vernacular terms, Sharad Joshi and the Shetkari Sanghatana became indisputably the most prominent strategic and ideological force among them.

Photo 1. Sharad Joshi addresses a crowd of Sanghatana participants in Vidarbha, 1997. Prior to the beginning of the rally, organizers asked all the women to be seated toward the front, under the shade of the tent canopy—a show of respect to the women in the movement.

As before, and now on a national stage, Joshi continued to articulate and expand upon the movement's ideological schema. He broadly defined the shetkaris the movement represented to include not just farmers and marginal peasants, but also agricultural laborers, rural artisans, and even the "rural refugees" inhabiting the streets of urban India—anyone whose economic livelihood is chiefly dependent on agriculture or *was* before the failure of the agricultural economy forced them out of it. He argued that the ultimate solution to the shetkaris' common problems was not access to land, but rather remuneration. Emphasizing that the movement stood for the uplift of *all* rural inhabitants—rural laborers as well as landowning producers—

Joshi insisted that higher prices would eventually trickle down to laborers as higher wages. To underscore this point, he encouraged all the movements in the ICC to include higher minimum wages for agricultural labor in their calculations of crop production costs, thus theoretically including fair labor wages within the target prices that they demanded.

Photo 2. Participants raise their fists at a Sanghatana rally during a call and response cheer. "Shetkari Sanghatana Zindabad! Sharad Joshi Zindabad!"

As the movement attracted greater attention, Joshi and the activists within the Sanghatana became renowned for speaking of their unity and their struggle in ways that often confounded urban observers. They declared themselves inhabitants of a country called Bharat, separate from India. They explained their movement as a struggle for independence (*swatantrya*) from the colonial rule of India, throwing off the yoke of the "black Britishers" (*kala ingraj*)—the Indian elite who inherited the governance of India upon independence from Britain. They described the future as a restoration of a mythological king, the demon Bali, long ago deposed by a brahmin dwarf who came begging for a parcel of land. One by one, as villages joined the movement, they erected signs declaring their village a province in the kingdom of Bali *(balirajya gav),* with the implication that it was off limits to "Indians."

In the midst of this massive growth, the Sanghatana raised the ire not only of then Prime Minister Indira Gandhi and her Congress Party but also the conventional left and opposition-party movements that had considered rural organizing to be their domain. The Congress Party as well as the left attempted to discredit the Sanghatana and its partner organizations in the ICC by declaring them "rich farmer" or "kulak" movements. They pointed to the presence of wealthier constituents, such as sugarcane growers, among their mass participants and attempted to rally rural poor and Dalit (untouchable) communities against the Sanghatana. In turn, the Sanghatana derided the Congress Party government as "looters" extracting value from the shetkaris to fuel the cities, and criticized the left parties as divisivists and beggars clamoring for subsidies and entitlements. Declaring "we don't want to beg, we just want the return for our sweat" *(bhik nako, have ghamace dam),* the Sanghatana continued to expand its base. It spread to new regions of the state. It encouraged shetkaris to grow new exotic or organic crops that enticed urban consumers and were not covered under existing price controls. It undertook local causes such as the construction of new roads for isolated villages, economic independence for women, regional independence for neglected regions of the state, the elimination of government corruption, and the forgiveness of shetkari debt. It opposed entrenched dominant-caste interests and conservative Hindu-ist and casteist rhetoric that served to factionalize rural social groups. In the process, it continued to attract agriculturalists with widely differing degrees of participation in the market, and reached out to rural laborers, women, and the low-status scheduled castes (SCs) that were historically marginalized by the politically dominant Marathas and ideologically dominant brahmins of the state.

The more time I spent with the Sanghatana, the clearer it seemed that the images of demon kings, god-like leaders, begging dwarves, separate countries, and the One Point Plan for higher prices were far more provisional and complex than mere elite attempts to beguile the masses into supporting a rich farmer agenda. Each of these was imbricated in complex fields of ideas and interests, and each represented, in one way or another, a wide range of rural concerns and subjectivities. But one question continued to bother me the most: if the movement truly represented a broad diversity of rural subjects, why did so many participants persist in reproducing the neatly bundled language and imagery of the movement's leaders despite pursuing their own often very different objectives? Why did peasants risk their lives in agitations, declaring, "Sharad Joshi is a god"? And why did so many marginal agriculturalists, including laborers who have little or no land, participate in the movement when it seemed to be centrally focused on higher prices for those who grow marketable crops? Was it enough to have some unguaranteed promise of these demands trickling down to them in some way in the future, or was something else going on?

Progressive solidification in the literature

One of the greatest challenges in social movement theory has been to understand the relationship between what people are *thinking* and what they are *doing* in the context of social mobilization. Some of the earliest theorizations of this relationship, often called "collective behavior theory," approached thought and collective action roughly in terms of the emotional impetus for action. Mass displays of dissent, in this view, were generally considered to be non-rational and often-spontaneous behavior stemming from shared frustrations (cf. Kornhauser 1959; Turner and Killian 1957; Park 1967; Lang and Lang 1961). In other words, mass social action arose from relatively uniform emotional reaction to shared circumstances (the *thinking* part), which, under the proper conditions, erupted into mass protest or rebellion (the *behaving* part). Collective behavior theory can be credited for attempting to uncover the genesis of collective action, but its take on the subjectivity of individual actors was essentializing and emotionalist.

Social movement theory in the later 1960s and 1970s viewed collective behavior with far greater political sensitivity. Applying the rubric of class, theorists made great strides toward better understanding the rational, material origins of individual and group behavior within a collective struggle. They also began to carefully examine movements' organizing processes and ideologies as tools for building and maintaining cultures of collective resistance (cf. Hinton 1966; Migdal 1975; Popkin 1979; Race 1972; Wolf 1969). This literature, relying heavily on the concept of class-consciousness, placed collective thought and collective behavior together in an ever-evolving relationship. Theorists began to recognize that organized dissent could be not only an *expression of* shared grievances, but also a *generative force of* collective identity and collective disenchantment. Although the collective behavior theory of the 1950s and 1960s and the classic class-based theory of the 1970s differed in important ways, both bodies of literature supported a similar implicit premise. This premise, which I have been calling progressive solidification, maintains that collective social dissent depends largely on broadly shared thought and identity. It implies, moreover, that as shared thought and identity expand, so do the potentialities for collective dissent to expand. This understanding of conditions and outcomes suggests that, by the time dissent has become a discernable and organized movement, participants have to generally agree in their perceptions of the world around them and their objectives for change. This is where theory fails to account for movements like the Shetkari Sanghatana. Under these theoretical dictates, multiclass, multi-interest movements such as the Sanghatana can only be understood as rational for *some* participants (those whose genuine class interests are in line with the dominant agenda and collective thought of the movement) but irrational for many others. This leaves no significant space within a movement for internal disagreement, differential interests, and differential identities.

Movement theory has at times offered interesting explanations for this solidarity. For instance, varied literatures have explained solidarity of thought by positing that participants are in the grip of "traditionalism" or "false consciousness" that prevents them from recognizing their true interests (for varied treatments on these themes see Moore 1966; Welch 1980; Mencher 1974). In both cases, participants' nonrational state of mind is generally accounted for by the influence of charismatic leaders and headmen and/or impersonal, superstructural, sociocultural forces that supersede and govern individual subjectivity.

These are extremely persistent ideas. The two most influential paradigms on consciousness and social mobilization leading to the turn of the twenty-first century continued to uphold the premise of progressive solidification in different ways. On one hand, there has been the "actor-centered," or "resistance" perspective in social theory, largely influenced by the work of political scientist James Scott (cf. Scott 1976, 1985, 1990; Prakash and Haynes 1991; Raheja and Gold 1994), This work, admirably, advocates what anthropologist Bernard Cohn (1987) has referred to as a "proctological" or bottom-up approach—that is, examining society from the perspective of its least enfranchised members. Resistance theory attempts to revalidate the acting individual as more than just a puppet of the elite. It argues that, despite all the weight of power-preserving ideologies and behavior-shaping institutions, individual subjects do, after all, have a consciousness of how the system works against them and how their interests are distinct from the interests that the system favors.

Resistance theory draws attention to ways hegemony fails to fool the weak—and the proof of this is in all the myriad, day-to-day ways that the weak resist power. Their resistance may be subtle most of the time, but it is resistance nonetheless. Resistance theory also points to cultural and societal structures that normally inhibit this resistance from turning into organized social protest by the weak.

In this way, resistance theory has helped to rescue subjects from charges that they are not conscious of their interests; it tells us that they are aware, but they are constrained in the ways they are able to act on that awareness. This literature was extremely influential for me during my time with the Sanghatana, and its influence will be seen throughout this book, especially in chapter 5. However, there are also ways that this literature plays into the social-scientific desire to see solidarity: it tends to romanticize a heroic underdog social group, glossing over heterogeneity, contradictions, and competition within the group (see critiques in Abu-Lughod 1990; Jeffery and Jeffery 1996: Prakash 1992).

Another very influential literature of recent decades has focused on ways that group consciousness becomes structured by ideational frames (cf. Snow et al, 1986; Benford and Snow, 2000; Poletta and Jasper, 2001). This work examines the social production of the ideas and narratives that inspire and mobilize people toward collective action. Frame theory points to methods and techniques used by empowered individuals and social groups to represent political ideologies in ways that have a desired effect with target audiences. Sociologists Robert Benford and David Snow (2000) call this "meaning work."

Frame theory has done a great deal to put the construction of movement meanings into the center of analysis, and I have relied on many of its conceptual contributions in my own analysis of the Shetkari Sanghatana—but the development of this literature thus far still falls into the traps of the solidification premise. The challenge to this literature is that it has an overwhelming tendency to view frames as effectively stable, monolithic productions of movement leaders. It offers slim space for dissent and disagreement from within mobilized groups that are associated with a particular frame, and even slimmer space for a group's less empowered participants to be coauthors of the frame or the meaning that it carries. For these reasons, sociologists Pamela Oliver and Hank Johnston (2000) have critiqued the current state of frame theory as an analysis of the "marketing strategies" of leaders—a perspective in which, as Oliver and

Johnston put it, the creation of meaning is primarily a "one-way, top-down process" rather than an "interactive negotiation of 'what's going on here'" (41).

As I will show in the case of the Shetkari Sanghatana, the struggle for signification occurs in many different ways and in different contexts. Ideas are not simply produced or consumed—rather, they are hotly contested in accordance with a wide range of subject positions. Individual subjects in the Sanghatana can themselves occupy numerous and contradictory positions. The result, as anthropologists John and Jean Comaroff have argued (1991), is that very few ideas have the "for granted" quality that could be characterized as truly hegemonic. Instead, most ideas, or systems of ideas that we can call an ideology, reside on a sliding scale in which they can be hegemonic to a greater or lesser extent, but never completely. Hegemony, as described by the Comaroffs, is something that is "always threatened by the vitality that remains in the forms of life it thwarts. It follows, then, that the hegemonic is constantly being made—and, by the same token, may be unmade. That is why it has been described as a process as much as a thing" (25).

The important thing to keep in mind in the context of a movement is that this sliding scale does not necessarily slide in any particular direction as a movement grows. At the same time, to say that hegemony is a process is not to say that it never has the appearance of stability. Even in the most broad-based movements there are contexts in which participants may experience a relative convergence of felt interests and social identities. These moments of convergence may be either fleeting or prominent in subjects' lives—but when they do occur they may outline what can be experienced *situationally* as an identity group.

The fragmented and integrated village

The question of shared and divergent interests in rural areas of the developing world has been a subject of ongoing scholarly debate. Scholars of India have engaged in this debate in the past by sparring over the precise number and types of rural social classes that exist in the country (cf. Brahme and Upadhyaya 1979; Silverman 1983; Omvedt 1981; Patnaik 1987; Thorner 1982). Where some saw "capitalist farmers," others saw "rich peasants;" where some saw "small farmers" others saw "middle peasants," and all debated about whether these categories did or did not include village merchants as a rural petty bourgeoisie. Even when scholars agreed that, for example, landless laborers were not farmers they vociferously disagreed on whether laborers constituted the "poor peasantry" or the "rural proletariat."

These were not mere debates about terminology. The point of these classifications was to identify the objective and material forms of existence that defined different experiences of power or oppression in rural society, and to theorize each group's genuine political interests. Unfortunately, these debates tended to be greatly abstract and removed from day-to-day rural Indian life. They generally failed to take account of the actual (rather than theoretical) subjectivity and experience of the people under examination, and they cast doubt on the legitimacy of other non-class forms of consciousness that contradicted or overlapped with the theoretician's abstract categories. These theorists tended to think of class as something that has a universal subjective experience. But, as political scientist Ashutosh Varshney has put it, class is not a "natural" form of subjectivity; the experience of one's class is shaped by many factors including such things as public policy, political mobilization, or the behavior of other people around them (1997). Class, in other words, is significant for understanding and assessing the logic behind agrarian mobilization—but other experiences and contours of subjectivity are significant as well.

Recognizing the need for a more nuanced understanding of subjectivity and interest, a number of scholars have sought approaches to socioeconomic stratification that are more sensitive to the multiple dimensions of power and subjectivity in South Asian settings. Some Indian scholars have discussed the need for a new rubric called "claste," attempting to develop a theoretical approach sensitive to both class and caste.[15] Other scholars, influenced by the pioneering work of Ranajit Guha (1983), have attempted to understand the complex interrelations of class, caste, and other structuring dimensions of ideological and material subjectivity through the even broader notion of *subalternity*.

All these debates on the most effective ways to analytically dissect and stratify village interests began to require much different nuance toward the end of the twentieth century. This was precipitated, in large part, by what scholars perceived to be a sudden rise in broad-based, multiclass, rural movements, such as the Sanghatana. For scholars, these movements demanded new approaches to thinking about political interest and identity—approaches that could reconcile social enactments of unity with social realities of differentiation.

Scholars have approached this reconciliation in essentially two different ways. A few have argued that rural unity can be a genuine expression of collective rural selfhood (Gupta, 1998; Lindberg, 1994; Omvedt, 1993, 1994; Varshney, 1995). This perspective does not deny rural social stratification or the diversity of rural interests, but recognizes rural unity as one of many situationally legitimate experiences of identity and interest structured by the economic and power differential between the country and the city. This view is summed up well by anthropologist Akhil Gupta (1998), who finds that agrarian unity is "not a singular subject position" but rather "a shared space of opposition" (97). A second and more populous group of scholars also views rural unity as a space of opposition between country and city, but emphasizes the efforts of empowered groups within rural society to create the illusion of unity for their own purposes. In this view, unity is more fabricated than real or legitimate. Political scientist Paul Brass, for example, sees the ideologies of India's broad-based agrarian movements as reactionary struggles that exploit the idea of "tradition" to help aspirational middle farmers compete for power against rural elites (1994a).[16] In Brass's view, these local interests attempt to create a semiotic world in which ruralism and nationhood are equivalent notions—or, as

Brass describes it, a discourse in which "nation = people = peasants = nature" (28).[17]

Brass is not alone on this. A number of scholars writing specifically about the Sanghatana have identified it as an ideological project of rich farmers or middle farmers, with little redeeming legitimacy for its poor and laboring participants (cf. Brass 1994a, 2000; Dhanagare 1994; Gupta 1997; Lenneberg 1988; Lindberg 1995; 1988; Rao 1996).[18] Dhanagare, arguing that the ideology of the Shetkari Sanghatana is a populist effort of Maharashtra's richest farmers over the middle and small producers, says, "if price is the driving issue, one must only look to see *who* has the marketable surplus" (72).

Despite apparent departures in the ways that scholars have theorized and classified the movement, they tend toward the same conclusion in relation to the pervasive premise of progressive solidification. Because the movement has endured, they tell us, its participants must feel a strong same-ness of interest and identity—regardless of whether that is authentic or manufactured by a specific interest group. As I will show, rural unity is neither wholly authentic nor wholly concocted—and participants in the Shetkari Sanghatana themselves recognize that this is a false dilemma. The shetkaris in the Shetkari Sanghatana recognize points of *con*vergence in their objectives and identities as well as points of *di*vergence. Moreover, they use both of these to their advantage.

As I will argue, the Shetkari Sanghatana's central ideology of a unified community ("Bharat", "the shetkaris") with a single shared objective (price) is more than simply an ideological ploy by movement leaders to craft a collectivity of participants. It is also exploited by its mass participants, who use it to contest rural social structure and promote their own objectives within the movement. In my view, rather than just reasoning backward from the ideology to identify a specific interest group that it appears to support (Dhanagare, 1994), we can understand the movement better by looking at its ideology as a product of complex material and ideological relations across the community as a whole, and exploring ways in which it is also expressive of subaltern experience and consciousness. This participatory perspective on ideological construction should not be conflated with an assertion of functional democracy or real equality within the movement—but it is very different than suggesting that one block of interests can dictate or mold identity and shared objectives unilaterally. As Akhil Gupta (1998) has argued, we cannot simply view rich farmer ideologies as co-optations of peasant experiential narratives. If we do so, we fail to acknowledge the discourses and productions of the weakest sections of the community who are its coauthors. This has implications not only for how we think about movements, but also how we understand ideological production and theorize leaders.[19]

From interpretive community to dialogical community

One way to think about a broad array of subjects coauthoring ideology and objectives within a social movement is to begin with the notion of an "interpretive community." This is an idea that I first encountered in the work of literary theorist Stanley Fish (1980). Fish argues that a group of people interacting together on a sustained basis eventually begins to attach substantially similar meanings to symbols that have currency within that group. This notion of a community of interpreters, emerging from mutual interaction with a pool of symbols, struck me as a useful point of departure for thinking about community and identity within a broad social movement.

In Fish's description, the subjects are a group of students in a class; the pool of symbols is a body of literature being used as classroom texts. As in other voluntary collectivities, students in a classroom share certain degrees of similarity that have brought them together. Presumably, in relation to the larger society, they have certain commonalities in terms of their background and experiences, their interests and objectives, and their overall cultural training. And yet, they are also individuals with distinct needs, desires, expectations, personalities, and social histories. In other words, the classroom is a resource that they utilize for both similar and dissimilar reasons, in order to satisfy interests and pursue objectives that are also both similar and dissimilar. To the extent that they constitute a community of practice, and a community of shared interests, it is logical and efficient for them to converge toward a community of shared interpretation. But the important thing to keep in mind is that this community is situational. The students diverge from one another in many ways. They live lives that are rich with other interests. They engage many other forms of community, in which they interact with other people, and through which they pursue objectives. Thus, for each student, this particular interpretive community is just one of many to which they belong. Each student's interpretive communities are situational, are contextually shifting, and, since they may call upon many of the same cultural symbols to communicate different meanings in different contexts, are overlapping or even contradictory.

There is another important point to keep in mind. Even within the specific interpretive community of the classroom, our students also use these same symbols to communicate their differences to each other (not just their similarities) and to pursue individual objectives (not just collective objectives). Thus, even within the community, the interpretive schema is not monolithic. Rather, it should be seen more in terms of a shared set of resources. Their shared interpretive schema is an idiom and collection of signs through which the students may contest, promote, and discuss variant meanings and objectives. This may often result in substantial interpretive consensus, but not *total* interpretive consensus. In these respects, participation in this hypothetical classroom is not unlike participation in the Shetkari Sanghatana: people participating in the movement and its idiom of community are not identical, are not consistently in agreement, are not unidimensional, and cannot be essentialized as fully "on-board" members of the movement in all aspects of their lives— nor even in all aspects of their participation. A better way to think of the movement collectivity would be as a *dialogical* community.

During my fieldwork with the Sanghatana, a number of innovations were emerging in social movement theory that support this perspective that participants' interpretive communities can be multiple, variable, and overlapping. The most exciting of these for me were those that looked at the creation of interpretive communities as an interactive and participatory process. What was thrilling was their suggestion that, while interpretive communities are shaped by power and unequal influence, they are not simply *established* by these forces. Even the least empowered participants bring new meaning to the community and may exploit the community symbols to pursue their own hermeneutic and socioeconomic objectives. Three theorists whose work I found particularly inspiring are sociologist Alberto Melucci, and anthropologists Gavin Smith and Akhil Gupta.

Italian sociologist Alberto Melucci (1983, 1989, 1996) treats social movements as an arena of symbolic confrontation. Similar to the literature on frames and framing processes, Melucci begins with questions such as "by what processes do actors construct their collective action?" and "how is the unity we observe in a collective phenomenon produced?" (1996, 382). But Melucci's work is more consistent in viewing a broad range of participants as interested and empowered agents in that constructive process. He argues that ultimate consensus on meanings and objectives is not a precondition for organizing or for a sustained collective effort. Rather, he suggests, social mobilization may occur in an ongoing context of disagreement and cultural negotiation as long as the form of that dialogical engagement allows both shared and divergent perceptions of interest to converge and form a relatively stable pattern of perceived benefits. Thus, Melucci's work leads us away from conceptions of social movement identities as bounded, monolithic, or solidified—arguing that, even in a mature movement, they contain within them sociocultural conflicts and fragmentation. Indeed, in Melucci's view, these are often the same conflicts and fragmentations from which a movement may be, in part, derived.

These aspects of negotiation and dialogical engagement are also central to anthropologist Gavin Smith's work on rural mobilization in Peru (1989 and 1991). Smith begins with the recognition that agrarian society is inevitably differentiated and heterogeneous. He argues that collective mobilization should be seen as the product of an ongoing and never complete dialogue among all of its participants, particularly with regard to the meaning of essential factors of routine daily life. His method is to examine actual dialogue on the central social norms of rural life, focusing especially on *key words* such as community, property, responsibility, and development.[20] This process can lead to a cultural collectivity akin to an interpretive community, but in Smith's work the community is always under construction and its attendant meanings are—to borrow the phrase from anthropologists John and Jean Comaroff—always dialogically engaged with the "forms of life" that they threaten (1991, 25).

Another work that I found influential during my fieldwork with the Sanghatana was anthropologist Akhil Gupta's (1998), writing about agriculture and peasant subjectivity in north India. Gupta sees subjectivity largely structured by dialogues—both within rural society itself, and across the experiential divide between rural and urban society. He argues that we need to closely examine both the form and the content of these rural discourses in order to fully appreciate their oppositional nature. Similar to Ranajit Guha's (1983) conception of subaltern practice as oppositional forms and content "made up not only of elements and tendencies that are in agreement but also those which clash and contrast" (334), Akhil Gupta suggests that any effort to identify "pure," uncomplicated forms of opposition or subjectivity is misguided. Instead, he argues that hybridity of thought, practice, and speech is the norm. So, labeling hybrid expressions as inauthentic might throw some critical light on intrusions of power in their construction but it also silences meaningful expressions of human experience.

These contributions are important for our understanding of the cultural processes underlying mobilization and identity. They help to flesh out and nuance our theorization of the agentive and subjective arenas in which individual agents ally with larger collectivities. They demand that we liberate ourselves from the idea that effective social movements are necessarily highly solidified or that their ongoing political significance depends on an ever-increasing solidification of objectives and identity. All of this impinges not only on how we theorize movements and participation, but also how we theorize their leaders. Are leaders, as the primary public ideologues of a movement, inevitably servants of their own personal interests or those of the class they represent? Can objectives and ideas expressed by leaders be partially authored and manipulated by less empowered participants? Or for that matter, can leaders themselves be co-created and exploited by these participants? These are questions that I will explore in chapter five.

Fieldwork: Bali's kingdom and its subjects

As I have suggested above, the Shetkari Sanghatana takes on a much different character when it is viewed from the perspective of the local level and through the actual practice of its participants. The closer we look at the movement, the more apparent it is that most theorizations about the solidarity of agendas and identities within a movement do not adequately apply to realities on the ground.

One of the reasons that large-scale social movements pose so many challenges to theory is that they are difficult phenomena to study. Due to their geographic spread and their often-high numbers of participants, social movements are most commonly approached by scholars through a macroscopic lens. Information sources for this macro view, as we might expect, lean heavily on speeches by the movement's key leaders and the movement's most visible performances of political action. The macro view also relies on newspaper accounts and press releases sent out by movement spokespeople. These data sources constitute what we could consider a complex of dominant, highly vocal transcripts on the movement—or something akin to what Indian sociologist

Andre Beteille (1991) has referred to figuratively as the "book view" of sociological phenomena. I do not by any means intend to condemn these sources. They can yield valuable insight into political and ideological trends, aid in the classification and comparison of social movements across multiple geographies, and contribute to an understanding of the broad mechanisms and strategies that underlie the successes and failures of individual movements. Indeed, I rely largely on such dominant transcripts at many points in this study. However, as I intend to show, a more ethnographically focused lens, attentive to the daily concerns and performances of actual individual participants, can offer a vastly different perspective on the movement and its inner logic. This is akin to the distinction that Indian anthropologist M. N. Srinivas (1996) has made between the "field view" and the book view. It is the job of the anthropological observer to draw these two views of large social phenomena together into an analysis that renders both classes of data compatible and mutually meaningful. As I intend to make clear, the book view, taken as social fact without the nuance of field data, is often skewed and misinterpreted to the point of dramatic misrepresentation.

Much of the difficulty in avoiding over reliance on these dominant transcripts stems from the fact that the size and complexity of many mass social movements make it difficult to conduct thorough and extensive fieldwork. This is further complicated by the fact that a movement may be in a continuous state of reinvention or have wholly passed into history before it becomes a subject of scholarly observation. My own field research on the Shetkari Sanghatana—a movement that was existing and observable but also vast and evolving—presented challenges to data collection that in some cases proved impossible to resolve. Nonetheless, I pursued the field view as an ideal. The resulting insights are no doubt only partially representative of the total phenomenon, but they are much more revelatory of the participant experience than could ever be possible through top-down sources.

The Shetkari Sanghatana is immense. It engages in activities on many different geographic scales, occurring at different times and in different places, across a state territory of over 300,000 square kilometers. At the time of my field research, its base of dependable activists included approximately 500,000 people, scattered in villages across the state.[21] As I have described earlier in this chapter, participation in the Sanghatana is physically observable because it employs mass, public actions to put pressure on social and governmental institutions. These actions include road and rail blockages, protest marches, crop withholding campaigns, *gav bundi* (closing a village to entry of government representatives), and massive rallies in opposition to government policy. Such activities attract participants in numbers from the hundreds to the hundreds of thousands. These tactics have led to police altercations and mass arrest, but have also won government concessions on many specific demands, particularly those concerning the government's procurement prices on important cash crops. But this is the movement in its macro view. As we peel the onion, we find much more happening. At regional and village levels, activists undertake many less prominent issues not inherently related to marketization and price. These include actions related to women's causes, anticorruption campaigns, debt forgiveness, the construction of new roads, or regional autonomy. Thus, the Sanghatana is not just one movement; it has many layers. The study of a social movement with these dimensions was extremely challenging. In my research I attempted to overcome the problem of scale by tacking back and forth between the book view and the field view—for example, between the centrally articulated agenda and local village issues; between statewide agitations and less prominent local actions; between accounts that are backed by the authority of significant leaders and the personal assessments of village informants.

When I went into the field in 1996, informed by the paradigm of progressive solidification, I had an overarching research objective in mind: to determine the processes through which the broad plurality of constituencies represented in the Shetkari

Sanghatana were able to construct a common political identity and a common consensus on the issues undertaken within the movement. About six months into my fieldwork, I began to recognize that this point of entry was flawed. In the initial months, whenever participants at the local level expressed movement-related goals and subjectivity that were not wholly consistent with the movement, I considered them to be outliers or more or less peripheral actors. In many ways, I disregarded their involvement as opportunistic and transitory. Although I realized that these actors must have some significance to the movement, it was not until after I had conducted many interviews in several villages that I began to recognize these participants as the type who constitute the critical base of mass involvement.

In the year and a half of fieldwork that followed that recognition, I began listening more carefully to the issues that motivated these opportunistic actors. I listened more carefully to the ways in which they spoke of their points of identity with the movement and how they utilized the symbols and performances of Sanghatana identity to express their own interests and subjectivity in different contexts—at mass rallies, in the village, and in the home. The more I compared the book view and the field view, the more I realized that an understanding of the Shetkari Sanghatana could not be acquired from either perspective in isolation. The learning resided in the areas of agreement and disagreement between the two.

Still, other challenges remained. Some of these I was able to overcome, and others not. One of these was the problem of braggadocio and inflated storytelling. Rural activists often describe the impact of local events and the numbers of participants that they were able to organize for events, in ways that are competitive and boastful—just like the proverbial fisherman, describing the one that got away in increasingly larger terms with every telling of the story. Although I did have some useful techniques for getting closer to the straight facts at organized events—conducting surveys, capturing events by camera and audio tape for later analysis, and using tricks like the "four-foot rule" for counting numbers of attending participants[22]—these techniques could only work in situations that I was able to personally observe. For others, whether they occurred during my stay in Maharashtra or some time in the past, I had to rely on a combination of newspaper accounts (preferably comparing the details offered in both "friendly" and "unfriendly" dailies), dependable informants, the Sanghatana's own official reports, and my informed gut instincts of what seemed likely to have been the case.

A final challenge that must be addressed is the question of representation. As mentioned, I made every effort to speak with people at different levels of the movement, from different class and caste backgrounds, in different villages, and in the context of different issues and movement-related activities. I pored through the Sanghatana's own archives, as well as external accounts. I traveled on my motorcycle to a wide range of rallies and agitations, and conducted interviews and observations in many dozens of villages where the Sanghatana is active throughout the state. For perspective, I interviewed leaders and participants in other Maharashtrian and Indian social movements. At some rallies and agitations, I conducted surveys with the help of field assistants in order to get the greatest possible cross section of participants' motivations and intentions. In the end, however, it is simply not possible to capture all that a movement of this scale and diversity represents. Therefore, while I believe the pages of this book are as true to another experience as I (an immersed but inevitably outside observer) could hope them to be, I also know that they are unavoidably partial.

A note on typologies, and the chapters ahead

There is no solid scholarly consensus on how to define a social movement, or on what characteristics are most important for differentiating types of movements. Disagreements on these become even more complicated when we fold in various perspectives on what is and is not a political act. At issue are some of the most basic aspects of the human experience: agency, subjectivity, identity, consciousness, and intent. I will consider these in depth over the course of this book. At this point, however, it is important to establish some basic outlines for my use of the term social movement in reference to the Sanghatana.

Most approaches to defining what is a social movement would agree on at least the following points: a social movement is a collection of people, informed of an ideology, cooperating over time to pursue change through some deliberate means. These are broad parameters, and they contain a number of different approaches and types of information that have been used to classify movements. Effectively, we can identify six categories of information that scholars have used for classification: *type(s) of participants* (class, caste, racial, ethnic, gender, religious, linguistic); *scale or spatial spread* (local, regional, national, rural, urban, spontaneous, sustained); *dominant issues* (political, economic, human rights, peace, ecology, equality); *nature of the desired change or ideology* (reactionary, revivalist, fundamentalist, reformist, revolutionary); *means of action or strategy* (violent, nonviolent, confrontational, persuasive), and *style of organizing or form of organization* (grassroots, formal, party-affiliated).[23]

In this book, I will be considering all of these categories of information. However, at all times I will be speaking of the Shetkari Sanghatana first and foremost as a *mass rural* social movement. By calling it a mass rural social movement, I mean to emphasize the Sanghatana's two most salient aspects in relation to the dimensions above: rural and mass. The Sanghatana is rural because the vast majority of its participants dwell in rural areas; its spatial spread is primarily confined to the countryside; and its dominant issues and desired change are focused on agriculture and/or village life. But while specifically rural, the Sanghatana exhibits a number of important characteristics that are also decidedly mass. Its people span class, caste, gender, and religion, and its ideology and issues have demonstrable appeal across this spectrum. Another mass characteristic is its geographic spread, covering an extensive territory across a very large state. Finally, it is also mass because it has a model of action that invites mass participation, and has an informal membership comprising individuals who are recruited into action in an informal and grassroots style.

This informal and grassroots nature contrasts sharply with some other types of rural movements in which participants are formally registered or enrolled members. These may include, for example, movements that are extensions of a political party, a union, or other formal membership organization. Such movements, more so than others, may seem to offer divergent counterpoints to many of the assertions that I will be making about the Shetkari Sanghatana. Certainly, I do not expect all types of social movements to conform to my assertions all the time. However, even when there are significant points of departure between one case and another, I do believe that the Sanghatana example sheds light on other forms of social movements.

Another important clarification at the outset is my use of descriptive labels referring to different types of Sanghatana participants. In general, I have tried to avoid two very widely used terms—peasant and farmer—as much as possible. My choice to avoid these derives, in part, from the deep theoretical morass that these terms conjure up. But my avoidance also reflects an appreciation for the ambiguity of the Marathi word *shetkari,* which best translates as neither peasant nor farmer, but as agriculturalist—a designation that embraces both of the English words peasant *and* farmer, as well as, to some extent, agricultural laborer. For this reason, I will most often use that somewhat clunky but appropriate word (agriculturalist), or the Marathi word itself.

When I do use more (seemingly) specific terms such as peasant or farmer, it is either in reference to someone else's usage (a quote, for example) or to clarify that I am referring to a distinctively positioned group that has been historically defined through one term or the other. When used in this way, I shall mean the term farmer to refer to those persons who are largely (but not necessarily exclusively) engaged in the business of agriculture *primarily* with intent to sell the produce of their own land in the commercial marketplace. These individuals may or may not work the land directly, but in either case generally employ labor as a component of their production process. Peasant, on the other hand, I shall mean to refer to individuals who are engaged in agriculture (on land that they own, rent, or otherwise use) principally (but not necessarily exclusively) for the purpose of household consumption and in-kind transactions, and who may also sell their labor to farmers. Able-bodied peasants (in other words those other than the elderly, the juvenile, or the infirm) and their family members are generally their own principal source of labor.

With these key understandings applicable throughout, the body of this study constitutes a progressive drilling down from the general to the particular—or from the book view to the field view. In the next chapter, I will consider the Maharashtrian context in which the Shetkari Sanghatana movement is situated. The chapter will begin with a geographic and historical overview of the state and its four major economic and historical regions. I will then look more closely at specific cultural and economic patterns across the state, including an examination of important castes and communities, with an emphasis on understanding some of the most significant terrains of social identity that are important to understanding collective identity and mobilization in the movement. The second half of chapter 2 comprises a fairly detailed analysis of several cultural and political phenomena in historical and contemporary Maharashtra that provide additional important context for understanding contemporary Maharashtra, and, more importantly, understanding mobilization and identity in the Sanghatana. This section is focused particularly on the rise and current significance of the Maratha Empire under the eighteenth-century king named Shivaji, the seven-hundred-year-old Maharashtrian bhakti movement called the Varkari Sampradaya, and the nineteenth century Satyashodhak movement of Mahatma Jyotirao Phule.

In chapter 3, I will dive more deeply into the Sanghatana itself, beginning with an analysis of the late twentieth-century economic landscape of rural Maharashtra. Building on the idea that at least a degree of convergent interests is necessary for an organized movement, this chapter will consider a number of possible reasons that a wide range of rural subjects—ranging from large landholders, to marginal landholders and even landless laborers—might find some measure of self-interest in the struggle for higher prices on agricultural produce. The chapter will then continue with an examination of the most overt level of the Sanghatana's ideology of price-based unity in the movement, considering a wide range of signs, strategies, and resources that the movement's central leaders are able to bring to bear in what I have called the "loud voice" representation of the movement.

Chapters 4 and 5 drill deeper into the view of the movement presented in chapter 3. Chapter 4 investigates the movement's representation of the One Point Plan for agricultural prices and the assumption that participation in the movement is steady, consistent, and fully understandable through that loud voice ideology and the issue of price. We will look at a wide range of other issues and programs that fall under the idioms of Bali's restoration and the One Point Plan, consider ways that these issues "trickle up" to leadership through participant input, and ways in which the depth and periodicity of movement participation are structured by this broader world of issues undertaken within the movement.

Chapter 5 is a continuation of this "field view" approach to the movement, shifting the focus away from issues and programs and considering in-depth the significance of Sharad Joshi's charismatic leadership and the movement-wide symbolism of Bali Raj. In this chapter, I will argue that the idiom of Bali and his mythological kingdom is deeply rooted in rural Maharashtra and provides a platform, within the movement, for dialogues on rural subjectivity and collective identity that challenge the cultural assumptions and socioeconomic dominance not only of the urban "black British" but also of the rural elite.

Finally, in the concluding chapter I will recap the main argument.

Notes

[1] *Zindabad*, meaning victory or long live, is a word of Persian origin commonly used in many Maharashtrian and India-wide social movement contexts. Given its Persian and Urdu associations, however, it is noticeably absent from movements that are expressly Maharashtrian nativist or Hindu nationalist.

[2] Given the centrality of this term in the following pages, I have chosen to depart from the use of italics for the remainder of the book. In this and other English transliterations of Marathi words I will follow a convention of anglicizing the construction of plural and possessive forms rather than using the more complicated Marathi constructions.

[3] When capitalized as Shetkari Sanghatana, or just Sanghatana, I am using these terms to refer to this movement organization and its proper name. When these terms are used separately without capitalization, they refer to their more generalized and common meanings of agriculturalist and union or organization, respectively.

[4] Formerly called Bombay in English.

[5] Demons and gods are generally considered in religious studies to be different types of supernatural beings. Although Bali is represented in classical Hindu mythology as a demon *(daitya* or *asura)*, I have here referred to him as god-like to allude to his frequent deification in popular discourse and ritual practice.

[6] *Bharat*, the Marathi and Hindi word for India, is not specifically or necessarily associated with Bali outside of the Shetkari Sanghatana movement. The word is widely used as a proper name for India when speaking in these or other Indian languages. In the Sanghatana's idiom, however, the English word *India* is always used in contradistinction to *Bharat* as a separate entity, regardless of the language spoken.

[7] In the resulting skirmish, one policeman was also killed.

[8] A pseudonym. In the following pages, I shall follow the convention of using pseudonyms for the names of most informants to protect their privacy. Exceptions to this include prominent movement leaders and public figures, including Sharad Joshi, when quoted in the context of a discussion that was mutually understood to be "for the record."

[9] Although in the following pages I will most frequently speak of Sanghatana objectives related to price as a demand for higher fixed prices, in the big picture this is merely an interim goal. The Sanghatana's

official long-run objective is a free economy in which prices would be determined by competition and market forces.

[10] This was the first noteworthy agitation staged by the emerging movement. However, some earlier discussions about launching a wider movement had already taken place during a minor agitation for the construction of an all-weather road serving villages in the area around the market town of Chakan.

[11] The APC, later to be reconfigured as the Commission for Agricultural Costs and Prices, or CACP.

[12] The fact that each delegate paid a nominal—but not insignificant—fee of 15 rupees to attend the three–day conference and help fund the Sanghatana's activities suggests a relatively deep level of commitment to the movement. The figure of 30,000 delegates is from S. C. Mhatre's introduction to Joshi (1988) and Sahasrabudhey (1989). Although I have no reason to question these sources, it is worth noting that Omvedt (1993) cites the figure of 18,000. Regarding the mass rally at the conclusion, again there is disagreement. Sahasrabudhey cites 300,000 and Omvedt cites 100,000.

[13] Joshi described a grassroots, nonbureaucratic vision of the movement that sounded more like a rapid reaction force. "We shall use guerilla tactics, and ensure that our agitations can be launched at forty-eight hour notice" (quoted in Omvedt 1993).

[14] Initially founded in 1985 as the Kisan Coordinating Committee (KCC), this was later changed to the ICC as rifts (both strategic and personal) developed between the Sanghatana and some of the other largest movements—most particularly the prominent Bharatiya Kisan Union under the leadership of Mahender Singh Tikait in Uttar Pradesh, and the Karnatak Rajya Ryot Sangha (KRRS) under the leadership of Professor Nanjundaswamy in Karnataka state.

[15] These emerging discussions were first described to me by the late Dr. Rajendra Vora, in the Department of Political Science at the University of Pune

[16] Conversely and unlike some other "new farmers movements" such as the KRRS in Karnataka, the Sanghatana's leaders wholly embrace scientific agriculture, including genetically engineered seeds, and they want more of it. They also embrace the advent of economic globalization and of Bharat's participation in the international marketplace. Its most vocal complaint is not about modernity, but about national and state policies that have inhibited the modernization and marketization of the

agrarian economy. But although the claim of "antimodernism" is not quite apt in the case of the Sanghatana, Brass and Nanda both brilliantly interrogate the uncomfortable similarities between many folk/nature narratives of agrarian unity and the neopopulist narrative of the militant Hindu right. Lindberg (1995), though less dismissive of the legitimacy of the "new farmers movements" has also explored what he calls their "ambiguous" relationship with nationalism.

[17] Prior to embarking upon fieldwork with the Sanghatana, I found some aspects of this position compelling. In a paper that I presented at the 22nd Annual Conference on South Asia, (Madison, Wisconsin, 1993) I suggested that the unity attained by the movement could be partially attributed to a broadly felt ambivalence toward India's historical and increasingly narrow technocratic approach to rural economic development. For an excellent critique of the "indigenous voice" angle in many arguments of this nature, see Gupta (1998) who finds similarity between scholarly efforts to find the "pure" and "indigenous" voice in oppositional practices and earlier reified constructions of "tradition."

[18] Dipankar Gupta (1997), who writes on the Bharatiya Kisan Union (BKU) in Uttar Pradesh does, far more than others in this camp, view peasant subjects as thoroughly rational actors who are not easily duped by ideologies that are against their real interests. However—similar to Brass and Nanda—he is deeply concerned about the supra-economic appeal of agrarian unity ideologies and their easy transference to populist communitarian and nationalistic movements.

[19] One interesting implication here, which I will address to some degree in chapter 5 and the concluding chapter, concerns how we theorize leaders who are held to be divine or semi-divine beings. Because of its idiom of the restoration of a "golden age" under king Bali, and the imputation by many participants that Sharad Joshi is an incarnation of Bali, the Sanghatana may seem to share a number of characteristics with movements that have been called "millenarian" or "revitalization" movements (cf. Adas, 1979; Wallace, 1956). As we shall see, however, the Sanghatana departs from the conventional theorization of such movements in significant ways. One of which is that participants within the movement vary substantially in how they characterize Sharad Joshi. Moreover, participants have actually been co-creators of Joshi's symbolic significance within the movement, and this co-creative activity is to a very large extent deliberate, rational, and strategic.

[20] Smith borrows the term "key words" from Raymond Williams (1976). In a book by that title, Williams traces transformation in the connotations of particular words he considers essential to social experience and argues that successive evolutions of these words correspond with the needs of the legitimizing ideology in historically successive relations of production.

[21] The 500,000 figure for village organizers is based on the estimates of leaders and participants interviewed. Although this figure is difficult to verify (the Sanghatana has no formal membership structure and, as we shall discuss later, the availability of most participants for organizational support waxes and wanes depending on the issues undertaken and personal circumstances), it does not conflict with my own assessment of the movement particularly during its high points of successful organizing.

[22] The "four-foot rule" is based on the assumption that rural Maharashtrians, when seated densely on the ground in a typical cross-legged posture, will consume an approximate average of four square feet per person. This insight became very helpful to me while measuring the attendance at several mass rallies. By pacing out the dimensions of a rally ground before participants arrived, it was then relatively easy to calculate the approximate number of attendees based on the proportion of the available space that was filled. My thanks go to S. C. Mhatre for enlightening me to this useful formula.

[23] Most definitions employ just a few of these dimensions. Here I am largely following Desrochers (1991), who outlines five categories but omits both organizing strategies and organizational structure, which are also widely used parameters for classifying movements.

2—Contours of community and place

Only in the mother tongue can speech and writing call up the deeply-felt imagery that people immediately understand. The language and ideas of this state—the language and ideas in which we must organize—are Marathi.

—A state-level Shetkari Sanghatana activist

My first visit to Maharashtra was in 1986, as a college student studying abroad. For six months of that year I was a guest in a humble, middle-class household in the dusty, noisy old commercial heart of Pune. My hosts were an older couple who were both Konkanastha Chitpawan brahmins—a specific and prominent but not large subcaste from the west of the state. It was in that household that I learned, through observation and conscious or unconscious imitation, what I understood to be the standards for "Maharashtrian" behavior. I picked up patterns of speech and vocabulary, gestures, rules for bodily use of space, etiquette, table manners, and other behaviors that helped me fit in, express myself appropriately, and avoid social gaffs in my new surroundings. Initially unbeknownst to me, I was adopting behaviors that were more specific than just Maharashtrian; they were also essentially urban, middle class, *Western* Maharashtrian (more precisely, *Puneri*[1]), and Konkani brahmin.

A decade later, during my fieldwork in Maharashtra in the late 1990s—after six years away from India, two years living among Rajputs and Jains in Rajasthan state, and a great deal of time spent with Maharashtrians from other regions, communities, and classes—Maharashtrian informants still occasionally teased me for eating, speaking, or gesturing like a brahmin from the western part of the state. "Ha! Look at how you position your fingers when you pick up your food...just like a Chitpawan brahmin!" "Where did you learn that word? That's what people say in the Desh...what *we* say is X." On other occasions, informants noted that I sat on my haunches like a shetkari or praised me for other behaviors that were more similar to their own, but my early socialization in Pune continued to be identifiable and richly meaningful to those around me. This experience helped me to better understand the depth and day-to-day expression of social variation within Maharashtra.

In this chapter I will begin to explore the range of identity contours and cultural meanings that are imbricated in the Shetkari Sanghatana movement, and the fluidity with which these may be expressed in varying contexts of social and political life. First, I will consider the idea of Maharashtra itself as a meaningful sociohistorical space, inhabited by a "Maharashtrian" collectivity. I will then outline some of the breadth and fluidity of social variation within this space. Finally, I will look closely at several examples of social organizing and collective contestations of meaning that can ultimately shed light on the phenomenon of the Shetkari Sanghatana. In the process, a number of recurring principles will be apparent. One of these is that identity contours within the state, ranging from the narrow strata of individual social groups to the very conception of "Maharashtrianess," are fluid and dynamic as much as they are historically embedded. Another principle is that identity groups often engage in substantial dialogue and achieve points of overlap and accommodation, even in cases when these groups are, in apparently fundamental ways, otherwise opposed.

When we closely consider the participant experience within the Shetkari Sanghatana, we see that most of its variable contours of identity and meaning actually have long established histories of expression and social functionality in other Maharashtrian contexts, past and present. Certainly there are aspects that are recognizably distinct and new about Sanghatana collective expression, but it is important to keep in mind that these expressions exist within and across a great diversity of parallel, intersecting, or even contradictory meanings and frameworks for identity in the state. What is new about the Sanghatana is its distinctive confluence of some possible meanings and the relative exclusion of others under the social rubric of the movement.

Expressions of common identity within a social movement, thus, might be best understood as intersections of interests, ideas, and opportunities that highlight *some* meanings and potential constructions of identity among mass adherents rather than a phenomenon that manufactures identity and meaning from new cloth. A movement provides contexts for the performance of this identity, and for the reinforcement of meanings that give the movement coherency.

Maharashtrian contours

The Shetkari Sanghatana is normally considered a Maharashtrian movement. This is essentially correct. Although it has made efforts to coordinate with other agrarian movements around the country, the Sanghatana's own targets of pressure tactics are predominantly Maharashtrian policies and administrators (rather than national); its geographic spread and organizational structures are, with few exceptions, confined to the state; and its ideological idiom is, in the words of the senior activist quoted at the top of this chapter, based in the "deeply-felt imagery" of the "language and ideas of the state." However, defining this state is complicated. As with most formal geographies, it is rich with both continuities and discontinuities.

The land that is today called Maharashtra encompasses the predominantly Marathi-speaking regions of west-central India. It is bounded in the west by the Arabian Sea, and along its land borders by the current states of Gujarati-speaking Gujarat, Hindi-speaking Madhya Pradesh, Telegu-speaking Andhra Pradesh, Kannada-speaking Karnataka, and Konkani- (and Marathi-)[2] speaking Goa. Along with the tiny state of Goa, it forms the southernmost part of India in which a Sanskritic language is spoken and it is often said to lie at the crossroads between the "wheat" culture of the north and the "rice" culture of the south.

With a rich history of shifting ties to its linguistically and culturally diverse neighbors, Maharashtra is what Indian anthropologist Irawati Karve (1968) has called a "culture contact region *par excellence."* A brief glance through the political history of the area further reveals much of this richness. Between the Aryan immigration into the region's Krishna and Godavari river valleys around 2000 to 1500 BC and the advent of Maharashtrian statehood in the mid-twentieth century, the region underwent tremendous territorial and political changes. Various parts of the Marathi-speaking land were ruled by the Ashokan empire from the north in the second century AD; by the Chalukya dynasty from the south in the sixth through eighth centuries; again from the north by the Yadavas; followed once again by a resumption of rule from the south by the reorganized Chalukyas in the tenth to twelfth centuries. By the onset of the fourteenth century, much of the Marathi territory was under control of the Delhi Sultanate along with a vast expanse of northern India—and, for a brief time, the Marathi area of Daulatabad became the capital territory of that empire under Sultan Mohammed Tughlaq in 1329. Thereafter, large portions of the Marathi-speaking lands were under shifting and contested authority of the Bahmani Sultanate and its various independent offspring (formed after Alauddin Bahman Shah broke away from the Delhi Sultanate in 1347), the Mughal Empire (which superseded the Sultanate in Delhi in 1526), the Marathi-speaking Marathas (who, under Shivaji Bhosale and his political heirs, extended their rule over much of the subcontinent starting in the late seventeenth century), and the brahmin peshwas (who, in 1749, wrested effective control of the Maratha Empire from within and presided over its continuing rise and eventual decline until finally being defeated by the British East India Company in 1819).[3]

During the European colonial era, most of the Marathi land was administered by the British crown under two separate multilinguistic states (the Marathi-Gujarati Bombay Presidency and, the Hindi-Telegu Central Provinces), but much of south-central Maharashtra remained under the formal rule of the Nizam of Hyderabad within the Marathi, Telegu, and Kanada-speaking Telangana province. Goa, with which coastal Maharashtrians share a close cultural and linguistic affinity, was under Portuguese control. Thus, for the last century and a half prior to statehood, the Marathi regions were administered under three different languages of state and from three culturally and geopolitically distinct urban capitals. The British ruled portions of the region from Mumbai, the up-and-coming center of maritime and industrial activity. Some areas were administered by the Urdu-speaking administration of the Nizam from Hyderabad, which was the capital of the Nizam's provinces. And, some areas were under Portuguese control from Goa.

This greatly simplified sweep of the Marathi land's history necessarily omits many smaller extensions of rule and political shifts over the centuries. My main point is that Maharashtra did not emerge as a geographically and politically unified land until well after India's national independence in 1947. The present state of Maharashtra was created in 1960—thirteen years after Indian independence—as part of the country's reorganization of legacy British and princely states throughout most of the country. Despite broad continuities of Marathi language and culture within its modern borders, the now politically unified state continues to be an area of diverse regional histories and subjective landscapes.

The modern state of Maharashtra is the third largest state in India. It is about the size of Poland, comprising 120,000 square miles (307,690 square kilometers) or approximately 9.36 percent of the Indian landmass (Dikshit, 1985). From east to west, it is geologically and meteorologically varied; however, with a few notable exceptions, most regions of Maharashtra lag behind national standards for agricultural productivity. As a whole, while the state accounts for 11 percent of the total national area in which food grains are cultivated, its contribution to India's overall agricultural production is just 7 percent (Sirsikar 1995). Agricultural economists in the state normally attribute this poor production performance to a number of factors. One is the low quality of soils around much of Maharashtra. Other factors include the often inadequate monsoon in the central portion of the state and the usually excessive monsoon in the coastal region; the state's relatively undeveloped irrigation capacity (only 13 percent of the gross cropped area is irrigated, as opposed to the national average of 30 percent); differential histories of land tenure and political participation across the state; and generally low reinvestment in agriculture. This poor performance in the agricultural sector contrasts strongly with Maharashtra's achievements in the industrial sector, in which it ranks first among all Indian states in per capita gross industrial output and industrial value added (Sirsikar 1995).

The singular and the heterogeneous state
Geographically, the most prominent division of Maharashtra state is that accomplished by the Sahyadri mountain range. On the west side of the mountains lies the long and narrow coastal strip known as the Konkan; on the east side lies the much larger inland portion—a high tableland that is geologically part of the central landmass of India called the Deccan Plateau.

The plateau land gives one the sense of an integrated natural space when contrasted with the coastal Konkan strip. Yet when considered independent of this contrast, the Deccan portion of the state manifests significant diversity of its own, divided by smaller ranges of hills and river valleys. Because of this, geographers conventionally speak of the natural and early cultural history of the state in terms of six different ecological zones. Starting from the west, the first of these is the coastal Konkan. On the plateau side, the five remaining zones are the Tapi river valley known as Khandesh, occupying the northern portion of the region now called Western Maharashtra; the Krishna river basin, covering most of the area historically called the Desh in the southern part of the Western Maharashtra region; the wide upper basin of the Godavari river, most of which constitutes the heart of the region called Marathwada; the Purna river valley, which runs between Marathwada and the region called Vidarbha; and the Wardha-Wainganga valley that constitutes the center of Vidarbha.

Over the span of history, these ecological zones have strongly influenced patterns of settlement, cultural interaction, and political developments in the area, and they underlie the current political, administrative, and popular division of the state into four contemporary regions. Each of these regions embodies important distinctions and histories that have implications for collective identity. These four formal regions—the Konkan, Western Maharashtra, Marathwada, and Vidarbha—are well understood by Maharashtrians to represent terrains of cultural, political, and economic significance as well as terrains of relative agricultural and industrial prosperity or "backwardness." Collectively, they comprise the thirty formal government districts of the state, each of which bears the name of the primary commercial and administrative town or city within its boundaries.[4] Because the Shetkari Sanghatana has experienced very different degrees of mobilization in each of these regions, it will be useful to explore some of the material differences that have helped define the regions well as some of the historic social, cultural, and economic experiences that have and have not been common to the regions over the span of Maharashtrian history.

On the western coast, the Konkan is the only region in the state that has remained virtually untouched by the Shetkari Sanghatana movement. The Konkan is a thin strip of coastland, stretching from the north of the state to the south, and nowhere reaching more than fifty miles in width. Although many parts of the region are among the most underdeveloped and inaccessible in the entire state, the Konkan is dominated in one form or another by the overwhelming fact of its principal district, Greater Mumbai, which is home to fully one-eighth of the total Maharashtra state population.[5] Because of the heavy Southwest monsoon along the coast—averaging a heavy 78 to 115 inches annually on the coastal plain and reaching well over 200 inches annually in the wettest part of the Sahyadri foothills—the Konkan appears verdant throughout most of the year. But this lushness masks its agricultural weakness. Poor soil quality and dampness make most parts of the Konkan strip inhospitable for the cultivation of cereals, oilseeds, cotton, and other crops that are highly valued in local and international markets (Dikshit, 1985). Thus, alongside primarily subsistence reliance on rice, fruit horticulture, and coastal fishing, most of the Konkan is dependent on what is often called the "money order economy" based on remittances sent from Mumbai by family members who have migrated to the city for formal or informal work (Sirsikar 1995).[6]

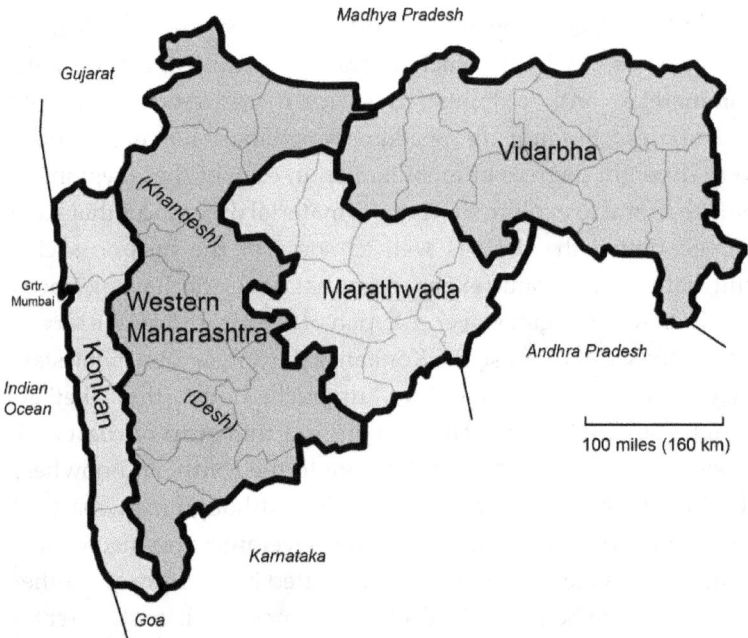

Map 1: Regions of Maharashtra

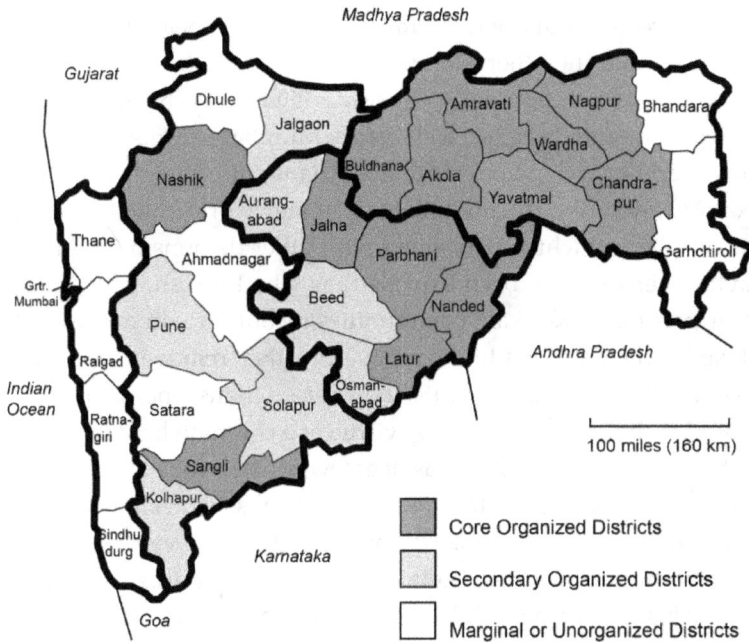

Map 2: Mobilization of the Shetkari Sanghatana by region and district, 1996–1998

This deep relationship with Mumbai can account for much of the distinctive contemporary character of the Konkan, including the highest literacy rates in the state (despite its weak rural economy) and the region's unusual rural voting pattern that is often as reflective of political trends in urban Mumbai as of issues related to rural Konkani development. During my fieldwork, this voting pattern was clearly evidenced in the 1996 Lok Sabha (House of the People) elections, in which the Shiv Sena party completely swept every district of the Konkan—the only region of the state in which it won every seat that it contested (*IE* 12/15/97).

It is significant here that the Shiv Sena, usually understood as a Maharashtrianist party of the Hindu Right and the Marathi Deccan heartland, originally established its following as a Mumbai-centric nativist movement aggressively opposed to wage labor migration into the city from outside of the immediate

Marathi region. But the Konkan's reputation for being somewhat separate from the other Marathi-speaking regions of the Deccan Plateau has been popularly acknowledged on both sides of the Sahyadri mountains since well before Mumbai became a place of any significance. Many Maharashtrians of the plateau, for example, describe the mythological nether world called *Patal* (a far off place to which Bali, along with other demons or forsaken mythic heroes, have been banished) as "the Konkan,"[7] while for their part many Konkani communities recount origin myths that attribute their lineage to places even further removed from the Deccan.[8] Nonetheless, despite these distinctions, the Konkan is united to the rest of the state by virtue of a common language and its historical role as both a coastal gateway and a coastal defensive bulwark for the Marathi heartland. The seventeenth-century Maratha king Shivaji, broadly touted as the first visionary of a unified Marathi domain, built several forts along this coast, and staged his coronation in 1674 at the hill fort of Raigad, strategically (and symbolically) overlooking both the coast and the plateau. The Chitpawan brahmin prime ministers who ruled over the same Maratha Empire in the eighteenth and early nineteenth centuries were themselves of Konkan stock.

Traveling from the Konkan eastward over the Sahyadri mountain passes, one descends from the foothills into the region known as Western Maharashtra which was, along with the Konkan, carved out of the Marathi-Gujarati province of the Bombay Presidency in 1960. Like Marathwada further to the east, much of Western Maharashtra lies in a rain shadow—a meteorological condition created as the atmospheric moisture of the western monsoon gets wrung out of the cloud cover as it passes over the Sahyadris and crosses inland over the first expanse of the plateau.

Western Maharashtra receives an average of only around twenty-seven to thirty-one inches of rain per year (Dikshit, 1986). In other words, while the Konkan receives far too much rain, Western Maharashtra receives far too little and several of its districts have been known since the earliest Marathi literature as areas of recurring famine (Karve, 1968). But this area also has extensive river systems flowing from the drenched slopes of the Sahyadris and, since the early twentieth century, this region has increasingly become the most heavily irrigated area of the state due to an extensive network of dams, canals, and expensive deep drill borewells.

Today, private and government sponsored investment in agriculture in this region is higher than in any other part of the state. Moreover, although the coastal area has the highest concentration of industry (in and around Mumbai), Western Maharashtra boasts the highest overall distribution of industry across its landscape. In these respects, Western Maharashtra is unparalleled in the state in the extent to which it has benefited from government-sponsored development schemes in industry as well as agriculture (Attwood 1992). Much of this industry is in the agro-processing sector, which is intimately (and politically) tied to direct producers of agricultural produce. This has an important bearing on the Sanghatana's up and down record of success in the region, and its ability to posit a fundamental divide between the interests of agriculturalists and their urban-industrial counterparts.

Politically and historically it is important to consider this region in terms of its northern and southern halves, which are known, respectively, as Khandesh and Desh. The Sanghatana's successes at organizing in both halves of the region have been decidedly mixed—nowhere near as steady and strong as in Vidarbha and Marathwada farther to the east, but a great deal better than in the Konkan. Its southern subregion, the Desh (literally, the country or the homeland) is in many ways the cultural and historic heart of the Marathi speaking region. It is the site of important temple centers for the Maharashtrian bhakti cults at Jejuri and Pandharpur, which have drawn waves of pilgrims from throughout Marathi speaking areas since medieval times. It is also the homeland of the two most beloved Marathi poets and devotional bhakti philosophers, the thirteenth-century saint Dnyaneshwar and the seventeenth-century saint Tukaram.

In the seventeenth century, the Desh was the scene of most of the Maratha king Shivaji's greatest battles and raids against the Mughal domains to the north, the Sultanates to the east and south, and various other claimants on the Marathi land. It is dotted with important forts taken or built by Shivaji and his political heirs during the early years of the Maratha struggle—and, hence, the Desh is the narrative setting for important tales about Marathi political liberation and state identity, signified through the image of the Maratha king. As we shall discuss, these dominant narratives on the Shivaji era, and the contours of identity they attempt to describe, are directly contested in the Sanghatana's invocation of Bali as a "people's king."

The Desh, more than any other region of the state, evokes the Maratha legacy. It continues to be the stronghold of Maratha politics in the state and, at the time of my field research, was considered the last bastion of (Maratha-controlled) Congress Party rule in Maharashtra. This is largely due to a complex system of political patronage, built over the years, based on party-supported investments in irrigation and party domination of the region's highly politicized cooperative sugar factories. Khandesh, by contrast, is more mixed both politically and agriculturally. Although it has seen some development of irrigation-intensive sugarcane production, this northern area is better known for its dryland production of onion and, in the western part of Khandesh, cotton. Prior to the stable political patronage of the twentieth century, however, both areas of Western Maharashtra were the scenes of shifting fortunes, mass relocations due to famine, and occasional revolts against local or external authority (Carter, 1988). In 1857, the northern part of the Desh and the southern section of Khandesh were both embroiled in one of British India's most significant agrarian uprisings, known as the Deccan Riots. This uprising, in many ways expressive of Maratha caste political aspirations against (non-Maratha) domestic agents of the colonial administration, posed a significant threat to the British government. It was instrumental in the passage of the Deccan Agriculturalists Relief Act of 1879. This act instituted important changes in the structure of agricultural taxes and levies, implemented controls on corruption and fraud, and abolished peasant imprisonment for debt (Charlesworth 1972).

The prosperous industrial city of Pune,[9] in the center of Western Maharashtra, is the second largest city in the state. It is widely regarded as Maharashtra's cultural capital, largely due to its Marathi literary and artistic contributions dating back to the brahmin peshwa era during which Pune served as the seat of the Chitpawan brahmin prime ministers for the Maratha Empire. Until recently, Pune's brahmin community played an especially disproportionate role in Maharashtra's institutionally recognized cultural production, but the latter decades of the twentieth century have supported a popular flowering of non-brahmin literature, theater, and visual arts. Prominent in this flowering has been literature from and about the Dalit community experience. Pune's status as an important site of Maharashtrian cultural production also extends from the importance of its social reform movements during the British period—such as the women's and lower-caste movement spearheaded by the nineteenth-century low-caste (*Mali*) reformer Jyotirao Phule—as well as the contributions of the city's intellectuals and organizers in the anticolonial movement and the later Maharashtrian statehood movement. The rural northern part of Pune district is the home of the Shetkari Sanghatana headquarters.

To the east of Western Maharashtra lie the historic regions known as Marathwada and Vidarbha, the two regions where the Shetkari Sanghatana has been the most consistently active. Marathwada was incorporated into the linguistic state of Maharashtra in 1960, at the same time as the other regions. Prior to that, from the sixteenth century onward, Marathwada was a Marathi-speaking region within the province of Telangana— predominantly under the Nizam of Hyderabad, but also for twelve years from 1948 until 1960 under the Indian flag when it was governed from Delhi.

Despite centuries of relatively stable but oppressive centralized rule from Urdu-speaking administrations in

Hyderabad, Marathwada stands out in its earlier centuries, as does much of Western Maharashtra, as an important cradle in the development of Marathi language and Marathi cultural identity. Many of Maharashtra's most important medieval sages and literary figures came from the region, including the still highly revered bhakti saint Eknath as well as king Shivaji's own guru Ramdas.[10] The Marathas fought repeatedly and with mixed success against the Nizam for control or containment of Hyderabad's Marathi territories. This continued until the British dethroned the Maratha Empire's last peshwa in 1819—but Marathwada was never successfully under Maratha control. Hence, in this region, as in most of Vidarbha, symbols such as Shivaji and the bhakti saints that have been used to narrate the collective "selfhood" and common aspirations of the Marathi people have held sometimes similar, but more often different, political significance compared with Western Maharashtra and the Konkan.

Under the Nizam's rule, this area was administered as part of Hyderabad's Telegu, Kannada, and Marathi-speaking Telangana region, where the majority of agriculturalists were deeply oppressed under feudatory *jagirdari* and *talukdari* systems of revenue (Brahme and Upadhyaya 1979). These were patterns of agricultural taxation in which individual holders of revenue titles over large tracts of land *(jagirdars)* or contract-based revenue collectors for the state *(talukdars)* wrung increasingly burdensome proportions of produce from the peasantry as tax. The arrangements were reformed in the latter part of the nineteenth century, but were replaced by a less direct form of state oppression via local landlords and moneylenders.[11] Land tenure and debt—as well as the ongoing issue of ethnic liberation and self-determination for the region—became the principal issues in the famous Telangana Uprising that overtook this area in the mid-twentieth century, beginning in 1946. At its peak, the uprising involved the participation of an estimated three million people across three thousand villages of Telangana. In some areas the movement continued for at least three years after the Nizam surrendered to the independent state of India in 1948 (Stree Shakti Sanghatana 1985).

Although most of Marathwada is covered with deep, fertile soil and lies within the state's major water drainage basin of the Godavari River, it has always lagged behind the other two plateau regions in its agricultural productivity. One reason for this is a natural shortage of rainfall resulting from its location in the rain shadow of the Sahyadris. By the time the monsoon reaches Marathwada, it is only able to produce about fifteen to twenty inches of rain per year (Dikshit 1985). Another reason is that, in much of the region, digging wells is technically and financially prohibitive because of the thickness and preponderance of bedrock. Although Marathwada should be able to make up for this loss by exploiting the surface waters of its extensive river system, the flow of water into the region is thwarted by the collection of government-built dams that are mostly located closer to the mountain slopes in Western Maharashtra. Many people in Marathwada are quick to point out ways such as this that the region is mistreated by the state. Many also attribute the region's underdeveloped condition to its long history of neglect and resource extraction by other distant rulers and their local administrators. Thus, although mass agitations in the region in favor of a merger with the other Marathi-speaking areas contributed to the success of the Maharashtra statehood movement (largely conducted under the banner of Shivaji's legacy), in recent years Marathwada has seen a popular movement for separate statehood based on claims of historical neglect by the state government and cultural distinctiveness from the rest of Maharashtra. The Shetkari Sanghatana has been very active in supporting this drive for separate statehood, and the Sanghatana's idiom of the internal colonization of Bharat is an effective signifier for this regional consciousness.

Vidarbha, the easternmost region of the state, is also a site of recurring calls for separate statehood—and is also supported in this by the Shetkari Sanghatana. With the bulk of Vidarbha farther removed from Western Maharashtra than is Marathwada, its regional characteristics are perhaps more distinct from those of the Marathi heartland. Although Marathwada receives the least share of the western monsoon, it is situated in the geographic center of the subcontinent, where it is fortunate to also receive a share of the other coastal monsoon blowing in from the east. Combined, the two monsoons produce an average annual rainfall ranging from about twenty-seven to nearly sixty inches across the region. This is at least double the amount received by the dry zones of Marathwada (Dikshit 1985). Thanks to this substantial rainfall and the thinner, less dominant bedrock that lies below its soil, Vidarbha has much better access to surface water and to underground water when wells are dug. Moreover, much of the region is graced with rich black soil—known locally as "cotton soil"—ideal for a number of important market crops, including cotton and wheat. These are precisely the types of market crops that are subject to the government levies and price controls opposed by the Sanghatana.

These ecological advantages have not led to equivalent agricultural prosperity in most of Vidarbha. This is most stark in the very easternmost districts of Bhandara and Ghadchiroli. These districts, which have high populations of *adivasi* communities (indigenous ethnic groups commonly referred to in Indian English as "tribals") nearing 40 percent of the total population in Gadchiroli district, have experienced distinct differences in their social and economic integration with the rest of the state (Singh 1995). This portion of Vidarbha has been an ongoing site of armed rebellion by the Maoist People's War Group (PWG, also known as the Naxalite movement) since independence, largely in response to government resistance to adivasis gathering forest produce or tilling land in state-owned forests. Although the Sanghatana has had substantial success organizing in most of Vidarbha, in the adivasi areas it has had as little success as it has had in the Konkan.

In most of Vidarbha, however, the Sanghatana has enjoyed a strong following and significant political influence. Much of this success can be attributed to conditions associated with the region's principal market crop—cotton. Unlike in the cane areas of Western Maharashtra, where production and milling have evolved into a comparatively symbiotic relationship between industrial sugar mills and producers, cotton growers in Vidarbha, Marathwada, and (where cultivated) Western Maharashtra are not as well enfranchised in regional formal politics. They are subject to a variety of marketing restrictions and price controls, and are substantially alienated from compensatory forms of political patronage and public investment. Here, the Sanghatana's call for market-equivalent pricing and the retention of profit within the rural community resonates with many local aspirations, as does the idiom of centralized black British domination.

The city of Nagpur, in the north-central Nagpur district of Vidarbha is noteworthy as the third largest city in Maharashtra. It is also the state's official second political capital (after Mumbai) and where the state government holds its winter session of the parliament. This two-capital model is a deliberate gesture toward political unity between the western and eastern sides of the state and has ameliorated some of the region's political isolation. It has helped to draw Vidarbha, formerly part of the colonially-constructed state of Central Provinces and Berar, more fully into the Maharashtrian fold. Another of the region's cities of distinction—by name at least—is Amaravati. According to myth, Amaravati was the name of the capital from which the demon king Bali ruled over his shetkari subjects until he was banished to the nether world of Patal.

Altogether, these four regions of Maharashtra contain about one-tenth of the total Indian population. About 70 percent of these people live in Maharashtra's nearly fifty thousand villages scattered across the state (*IE* 12/20/98).[12] These villages show tremendous variation. There are significant differences between them in terms of their natural endowments, their access to public services and state-sponsored infrastructure, and their local economic and political interests. For example, at the time of my fieldwork, only thirty thousand of these villages (about 60 percent) were connected to each other and to market centers by paved roads, and over 3,000 of them had no road connections whatsoever, either paved or unpaved. Of the villages that had zero road connections, two-thirds of these (2,047) were in Vidarbha, and a significant remainder of them were in Marathwada (ibid.). Such uneven patterns of development across the state—and unequal access to the political and financial power of Pune and Mumbai that these cities represent—have continued to fan the flames of calls for separate statehood for Marathwada and Vidarbha, the two regions in which the Sanghatana has had its strongest following.

As with separatist aspirations in other parts of the world, it is easy to discern certain interests of elite stakeholders in the push for political independence. In many ways, the separatist demands can be seen as a strategy of regional politicians and industrialists who feel overshadowed by the political and industrial dominance of Western Maharashtra and Mumbai (Mudholkar and Vora, 1984). But separation also gives voice to nonelite interests and frustrations. These include popular desires for a more responsive government, concerns about economic surplus in Marathwada and Vidarbha being used to fuel development in the state's west, and desires for greater returns on cotton, the principal cash-earning crop in these regions. Coterminous with this sense of political and economic alienation are important social differences between the eastern and western halves of the state that help explain the popular interest in separation from Maharashtra. For example, Marathwada and Vidarbha are the two areas of the state with the largest Dalit and Muslim constituencies. These groups often feel under-represented—politically and culturally—in the larger state. Even among the Marathas and Maratha-like castes of these regions, historical cultural identity with the politically dominant Marathas of Western Maharashtra has been notably weak (Sirsikar 1995).

At the same time, however, Marathwada and Vidarbha and most of their rural communities (with the primary arguable exception of the isolated adivasi communities) also participate in a distinctively Maharashtrian popular culture, based in shared language and shared cultural symbols that are distinctively Marathi. Thus, "Marathi" as an ethnic identity, and "Maharashtra" or any of the Maharashtrian regions as specific geographies of identity, are historical and symbolic constructs. As with other terrains of identity, these emerge from unique histories of experience and are shaped by the interplay of variously situated interests and competing representations of selfhood. At different times and in different contexts, the Maharashtrian terrain of identity and interest can serve as a popularly functional basis for shared community and sentiment. In other contexts, it may be trumped by—or it may intersect with—other identities such as those of tenants, agriculturalists, industrialists, members of a particular caste or community, or inhabitants of a geographic subregion.

Communities of Maharashtra: continuity and dynamism
What is apparent then is that despite the discernable, overarching Maharashtrian-ness of the historical, cultural, and behavioral patterns found in the Marathi-speaking land, the specific identities that Maharashtrians experience and perform may shift from context to context. My own experience as an unwitting performer of coded behaviors, described at the opening of this chapter, helped me understand some of the ways that sameness and otherness may be expressed and interpreted in daily life. Within each region, there are noticeable variations in dialect, castes and communities, and patterns of village settlement. All of these, just like the varied political, economic, and cultural experiences across the region, constitute diverse lenses of experience through which ideologies of political and cultural sameness or "otherness" may be interpreted in any given context. These contextually shifting contours of identity are equally significant at the village level, where speech patterns, worldview, and cultural conventions for interacting with the mundane and the sacred can vary substantially across a single village. In any given context, these variations may appear subtle, fitting comfortably as pluralist cultural expressions within the overall

cultural unity of the village, the region, or the state. In other contexts they may be overt and deliberately reinforced through oppositional practices that assert or contest relative lines of status, power, and selfhood. All of this impinges on the potential for a mass movement to "fix" a particular identity in the consciousness of its participants.

The social and geographic communities in which people are enculturated and participate are very much *behavioral* collectivities. Subtle behavioral indicators are always present in social interaction, in greater or lesser degrees of conscious, purposeful performance. It is the specific *context* and meaning of any social interaction that determines which aspects of behavior an actor will intentionally perform, and which will be interpreted by the interacting individuals as socially salient in that context. As a result, interactions across behavioral groups will either tend to reinforce and demarcate boundaries within the community of interacting subjects, or push them toward a negotiated shared space that emphasizes what they do have in common. This gives us some insight into phenomena like the Sanghatana, and suggests that a major challenge for a mass social movement is to create or facilitate contexts of collective action that encourage performances and behavioral interpretations that support the movement collectivity, rather than those that emphasize fragments and multiple boundaries. Religion, caste, class, and region are just a few of the possible boundary lines that can be described through casual, unconscious behavior. But while these boundaries can be extremely significant in many contexts of social interaction, in other contexts they are more fluid and variable than is often recognized.

Religious community represents a social boundary that in recent years has been highly contentious in Indian public life and South Asian scholarship. If we approach Maharashtra by the standards of inclusion-exclusion used by the Indian census bureau, we could say that the population of Maharashtra is approximately 81 percent Hindu.[13] The same source tells us that Muslims number just over 9 percent in the state, with this proportion varying somewhat from district to district. Buddhists, who are almost entirely of the *Mahar* Dalit community that converted to Buddhism from Hinduism in the mid-twentieth century, represent just over 6 percent of the population. The remaining 4 percent includes Christians, Jains, Sikhs, Zoroastrians and Jews (Census of India 1991). Numerically and subjectively—in very many contemporary contexts—it is an overwhelmingly Hindu society.

In daily interaction, however, these religious boundaries are more difficult to determine. A great deal has been written about the historical and politicized construction of religious boundaries—particularly with regard to the terms Hindu and Hinduism—that attempts to encompass a broad set of practices and beliefs within a single shared social border (cf. Inden 1990; Laine 2003; Lele 1995; Oberoi 1994; contributions to Richman 2000).[14] In actual social practice, the context of interaction and the potential rewards of any particular self-identification are often more important than the formal boundaries ascribed by outsiders. Outside of some political and public ritual contexts, Hindus may interact selectively among themselves (segregating themselves from other Hindus), or may interact with members of other religious groups in equally meaningful convergences of interest, such as shared regional or economic pursuits. Even among Hindus, daily interaction is marked by variable contexts of meaning and interest that foster other shifting or competing identities, such as caste identities.

The principal castes of Maharashtra include the various communities of brahmins who have wielded disproportionate ideological control in the state, and who collectively number just about 3.5 percent of the total population (Karve 1968). This relatively small percentage has been widely cited over the years as "the rule of the three and a half percent" by Maharashtra's Dalits ("untouchables"), who have struggled for a voice and subjectivity unstructured by cultural constructs of brahminical caste hierarchy, brahminical proximity to the divine, and brahminical purity. These Dalit groups, especially the *Mahar*, the

Mang and the *Chambhar* communities, constitute about 10 percent of the population across the regions and are particularly well organized in the southern districts of Marathwada (Dikshit 1985; Sirsikar 1995). Other communities that have historically been marginalized include the adivasi groups, representing about 5 percent of the total population (Dikshit 1985). These include the *Gonds* (in far eastern Vidarbha), the *Warlis, Katkaris, Kolis,* and *Thakurs* (concentrated in the hill areas of northern Konkan and central Western Maharashtra, representing up to 40 percent of the population in some districts—as is also the case with the Gond community in Vidarbha), and the *Bhils* (in the very north of Western Maharashtra).

Numerically, the most prominent caste group in the state is what is commonly referred to by social scientists as the *Maratha-kunbi* caste cluster. This group, which represents about half of the population across the state (Karve 1968) and has its highest concentrations in Western Maharashtra and Marathwada, has generally been considered what anthropologist M. N. Srinivas (1987; 1996) calls a dominant caste.[15] The group is associated with the Maratha Empire from the seventeenth to the early nineteenth centuries, and rose to electoral prominence in Maharashtra after Indian independence in 1947.

This Maratha-kunbi cluster is demonstrative of ways in which constructions of identity and unity in the state are normatively shifting, contextual, and responsive to changing opportunities afforded by different identities in different situations. Although there are many reasons for the conceptual utility of this broad cluster of castes—which can include Marathas proper, as well as *Malis, Telis, Lohars,* and *Dhangars* among others—it is also extremely problematic. Caste in Maharashtra, as elsewhere in India and as with other forms of identity, is in large measure maintained through a constellation of popular signs of identity and oppositional constructs. But, similar to regional, religious, and ethnic identities, these oppositional boundaries occupy numerous different levels of abstraction, depending on the context and the interests involved. Thus, within the Maratha-kunbi cluster (as with other composite caste groups), Marathas and the kunbi castes vacillate situationally from unity to division.

Within the Maratha-kunbi cluster, much of this vacillation is related to one of the very unusual aspects of caste in Maharashtra compared with most other states: unlike most Indian states, Maharashtra did not historically have an indigenous community of *kshatriyas*—the caste that, under the fourfold ideology of *varnas*, is entitled to political rulership.[16] As I have already briefly discussed, up until the Maratha Empire, political rule in Maharashtra came predominantly from outside of the Marathi-speaking community, but was supported by local agents and sanctioned by the ideological rule of Maharashtra's brahmins. Thus, prior to the Maratha rule of Shivaji in the seventeenth century, and the brahminical conferral of kshatriya status on his (previously shudra) Maratha clan, Maharashtrian society was much more sharply divided between the ideologically (and, indirectly, politically) dominant brahmins, on the one hand, and the shudra peasant masses, on the other. The latter were collectively known as kunbis—not a caste designation, per se, but a reference to "people who work the land"—and this category included the Marathas, as an agrarian caste.

When Shivaji's clan was lifted to kshatriya status, other clans of Maratha commoners and the kunbi masses were in some respects lifted in their social status as well. But not all agricultural communities rose equally. The ascendance of Shivaji's clan introduced new dynamics in the fractionalization of the agrarian castes across the ideological kshatriya-shudra spectrum. In some contexts, Maharashtrians look at the divide between these groups as sacrosanct. In other contexts, it is considerably more fluid. For example, Maratha political elites who have a clear interest in forging a mass middle-caste identity that is responsive to its own interests frequently invoke common cultural patterns of the agrarian castes in order to represent the Maratha-kunbi cluster as a "natural" and united constituency of interests. Shudra kunbis, for their part, often embrace this representation in order to ally themselves with Maratha political and social capital when it suits their needs. In addition, the kunbi community's identity alliance with Marathas, and its adoption of behaviors that are popularly understood to signify Maratha status, is a time honored strategy for kunbi families and kunbi caste groups to improve their social position or seek legitimacy for new economic and social roles—much in the way that Shivaji's clan upped its own ranking, in part, by the adoption of kshatriya behaviors imitated from the Rajputs of Rajasthan (Laine 2003). This aspiration appears to have been deeply embedded in Maharashtrian society for several hundred years. It is reflected in a common Marathi aphorism—"When a kunbi prospers, he becomes a Maratha." This is sometimes uttered in disparagement of status-seeking families, but other times as a more matter-of-fact commentary on social reality. Something along the lines of "that's just the way the world is."

Despite claims of similarity and shared identity that are often made by both Marathas and kunbis, significant distinctions between kshatriya Marathas and the shudra Maratha-kunbis are crafted and maintained in both overt and subtle ways. *Assal*—or true—Maratha families reinforce their kshatriya status by tracing their lineage to ruling Maratha families of Baroda, Gwalior, Satara, and Kolhapur; they align their true Maratha (as opposed to *common* Maratha or kunbi) status with historic family entitlements as landlords and as owners of land-grant jagirs[17] under the Maratha Empire. Historically, they have most commonly served as village *patils*, or village heads—a position of continuing symbolic if not formal authority throughout the state (O'Hanlon 1985). Ordinary kunbis cannot claim these hereditary achievements.

These status claims in relation to other "mass middle" Maratha-kunbi castes are reinforced today through a wide range of daily practices and social proscriptions. One important window on this is rules of endogamy and hypergamy. Assal Maratha families have conventionally permitted daughters of relatively well-off kunbi families to marry "up" into relatively poor Maratha families, but have not conventionally given their own daughters for marriage "downward" into common Maratha or kunbi families (Karve 1968). In addition, boundaries between these caste groups are symbolically drawn through distinctions in the dress, speech patterns, and ritual practices of each group, and the social commentaries that each group directs toward the other. The Dhangar community, for example—sometimes considered to be within the cluster and sometimes not—is closely associated with worship of the god *Khandoba*, who is typically the central household deity of Dhangar families. The spiritual practice of the Khandoba cult is based in a distinctive type of worship known as *sakama bhakti*, a devotional form in which god is expected to grant wishes to the devotee. Marathas and other upper castes often consider this form of devotion crude and superstitious. Similarly, ritual practices associated with the demon Bali are highly variable across village castes, and suggest direct contestation of Bali's meaning between those who work the land (kunbis) and those who exercise authority (large landholders, patils, and other elite Marathas). I will return to some of these ritual elements and meanings later, particularly in chapter 5.

The important thing to understand at this stage is that these various points of identity are situational rather than hegemonic or monolithic. This can be seen in the way that individuals within the so-called Maratha-kunbi cluster allude to these identities in interviews and routine conversation. For example, in one village that was often described by informants as overwhelmingly Maratha, I found that some elite Maratha families might, in one context, tell me "*everyone* in this village is Maratha." In other contexts, however, the same individuals derided certain families, their social behaviors, or their ritual practices as being low-caste (*khalchi* or "from below"). Likewise, I found that many common Marathas or kunbis in some contexts used the term Maratha for just about any native speaker of the Marathi language (including people from what would normally be considered non-Maratha or non-kunbi castes). In other contexts, they used Maratha as a (self-inclusive) term for people who were neither brahmins nor Dalits. In still others, they used it as a term of disparagement for someone whose behavior is perceived as high-handed or elitist—"that fellow is such a Maratha!" In my experience, the specific vocabulary and contexts of such talk vary from village to village, and are undoubtedly related to a village's caste composition, caste relations, settlement patterns, and the local and regional structures of power that influence village life. Generally, however, it seems fair to say that there is a clear pattern of situational fluidity in the use of some caste identifiers— particularly with regard to the large percentage of the population that is the agrarian Marathas and their associated castes.[18]

The Maratha-kunbi cluster, then, is a highly negotiated and contextual identity group. And although an inclusivist notion of Maharashtra as the "land of the Marathas" is one of the dominant tropes of the Maratha political elite through which Marathi nationalism has been narrated, the spirit of Maratha-kunbi unity varies significantly across the state and from context to context. In the strictly formal political context, unity within the Maratha-kunbi cluster is today most powerful only in the prosperous cane-growing districts of Western Maharashtra, where it is closely tied to formal political and economic affirmation of this group through the state-run cooperative sugar mills and associated political patronage. There, Maratha-kunbi political unity has been structured as much by the sugar cooperatives, and the political opportunities that these represent in a Maratha-dominated political region, as by narratives on the virtues of Maratha-ness and the role of the Maratha Empire in the formation of the state. Moreover, the blurring of social boundaries within the cluster seems to run particularly deep in the region's history. This is probably attributable to political upsets (prior to Maratha stability in the eighteenth century), famines, and migrations over many generations prior to the twentieth century that induced social reorganizations and changing hierarchies of status in many Western Maharashtrian village communities (Attwood 1988). This strong construction of Maratha-kunbi unity, combined with the beneficial ties that agrarian Maratha-kunbis have established with industrial and political elites, offers some insights into the

Sanghatana's relative inability to mobilize these districts with anti-casteism and anti-statism messaging.

In the eastern regions, the Maratha-kunbi cluster does not exist as a functioning unity as much as it has in the past, or does in the west. Russian anthropologist Irina Efremova (1997), in a broad survey of caste conservatism in the Maharashtrian regions, has found that endogamy and other boundary-crafting practices among the individual Maratha and kunbi subcastes are more rigid in Marathwada and Vidarbha than elsewhere in the state. In other words, these regions exhibit greater resistance to the unity of Marathas and kunbis as a clustered identity. Today, this may be partially attributed to the relative weakness of structures of Maratha patronage, which are not as rewarding for the agrarian mass-middle in these developmentally neglected regions.[19] Historically, this may reflect the isolation of these regions from the fortunes and patronage of the Maratha Empire, and their relative social and political stability under other regimes.

But religious community and caste are certainly not the only dimensions of experience in which identity might be expressed or performed differently in changing contexts of social interaction. One dimension of great significance is an individual's or a group's relative power and economic position. A critical aspect of this is one's position in relation to production and the market. In this, the Maharashtrian pattern is distinctive because Maharashtra rural *class* mobility has historically been more fluid than in most other Indian states (Omvedt 1994a). In other words, a caste, or a family from a caste, may change its position in relation to the means of production relatively more easily in Maharashtra than in other states, even though opportunities to change its caste ranking may vary from place to place within the state. This helps to explain why some sections of the rural community may seek representation through a caste-affiliated interest group (for example, the Dalit-issues Republican Party of India, or the Maratha section of the Congress Party) in some places and contexts, and a caste-transcendent agrarian issues movement (for example, the Shetkari Sanghatana or a rural labor organization) in other contexts.

Another distinctive feature of Maharashtrian rural economic life is the high degree to which broad cross sections of castes and classes participate to some extent in the market. In most Indian states, small and marginal plot holders tend to be subsistence-oriented in their food production and rely on selling their labor when they need cash. In Maharashtra, however, small and marginal producers share some affinity with the class that political scientists Lloyd and Susanne Rudolph (1987) characterize as "bullock capitalists" or what sociologist Dipankar Gupta (1997) has described as "farmer-peasants"—agriculturalists who have limited material resources and limited opportunities but still produce some salable crops and interact with the market. This broadly caste-transcendent status as producers and market participants (or what might be called farmers) represents a very significant potential contour of situational identity. This may be especially the case in contexts where caste is a comparatively ineffectual affiliation for pursuing economic and political objectives. Moreover, this pattern of broadly distributed market-oriented production in Maharashtra has been shown to be increasing since the latter decades of the twentieth century (Lenneberg 1988; Omvedt 1994a; Varshney 1995). This helps to explain the possibility of the Sanghatana in the last decades of the century as a price-oriented movement that cuts across communal and caste constituencies.

As the foregoing suggests, the parameters for social identity in Maharashtra are clearly not fixed to strict social lines of division, nor do they remain constant as circumstances and opportunities of interaction change from one context to another. On the contrary, all of these social boundaries and identity constructions are, to take some liberty with anthropologist Nicholas Dirks' (1992b) phrase, "castes of mind"—shaped not only by the determinations (and indeterminations) of others, but also by the perceptions and actions of subjects claiming affiliation with any group. Following the insight of Pierre Bourdieu (1990), the social practice of these affinities as a *habitus* of thought and behavior cannot be wholly understood either as the mechanical constructs of elite power or as the fully voluntary and individualist affiliations of subjects. They are produced and maintained through the interplay of both. But it is also important to recognize that these contours of identity are plural and not mutually exclusive; rather, they are shifting and often overlapping responses to different contexts of opportunity and different logical levels of action and meaning.

Signifying mass identity in Maharashtra

As we have already seen in other contexts above, popular and oppositional conceptualizations of situational unity across large cross sections of social groups in Maharashtra are not unique to the Shetkari Sanghatana. While the Sanghatana and its participants have crafted a novel constellation of ideas and contexts of action, they have had to construct these in accommodation or contestation of other perceived terrains of collective interest or corporate identity. Just as in the construction of communal, caste, class, regional, and statewide affinities, the Sanghatana's own propositions of shetkari unity, of Bharat, and of the ideal rule under Bali Raj (the Bali *realm*) have had to contend with other contextual and historically deep constructions that may contradict, fragment, or transverse these categories. As a mass movement intended to foster a socially functional agrarian identity across a broad cross section of rural society, the Sanghatana has also had to contend with other coexisting and historical constructions of identity that are signified through many of the same popular cultural resources.

Through its history, the Sanghatana has made efforts to organize in all the Maharashtrian regions (though, to a lesser extent in the Konkan) and across all Maharashtrian rural social groups. Given its support of regional secession movements and its conceptual division between urban-industrial Maharashtra and rural Maharashtra, the Sanghatana cannot be readily equated with the Marathi nationalism of organizations like the Shiv Sena.

At the same time, however, its attempts to define broad contours of unified interest through language, ideas, and symbols that are deeply Marathi forces it to challenge or affirm other ideologies and cultural signs of collective selfhood through which the evolution of the Marathi nation has been narrated. In the remainder of this chapter, I will examine three cultural phenomena that are deeply implicated in popular Maharashtrian narratives of "the people." These are the Varkari Sampradaya pilgrimage movement, king Shivaji's Maratha Empire, and the Nonbrahmin Movement. In subsequent chapters, I will point to some of the important ways that meaning work in the Shetkari Sanghatana addresses these pools of popular signification.

Mass community and the Varkari Sampradaya
One of the most prominent and distinctively Maharashtrian cultural phenomena in the state is a pilgrimage cult known as the *Varkari Sampradaya.* Called "the great Maharashtrian pilgrimage" by anthropologist Victor Turner (1974 [1973]), it is one of the earliest recorded collective movements in Maharashtra. The history of the movement stretches back more than seven hundred years, and today it is unquestionably the most widely practiced pilgrimage movement in the state.

The Varkari Sampradaya (meaning literally the "path" of "one who makes the walk") emerged in Maharashtra sometime between the late twelfth and the early thirteenth centuries. It is rooted in an ancient transcendental devotional philosophy called bhakti.[20] One reason for its importance in understanding modern Marathi identity is that the movement is generally credited with several pivotal historical contributions in the flowering of Marathi cultural consciousness. Among these are the legitimization of the Marathi vernacular as a literary and philosophical language; its establishment of a platform for cross-caste social unity in the collective practice of the cult; and, through the annual pilgrimage to the small town of Pandharpur in the central Desh, the encouragement of cultural, intellectual, and material exchange throughout the Marathi-speaking regions (Karve, 1968; Sardar, 1969; Zelliot, 1982, 1987). Today, the movement is so embedded in the lore and performative culture of the state's regions that Maharashtrian anthropologist Irawati Karve (1988) once remarked "I found a new definition for Maharashtra: the land whose people go to Pandharpur for pilgrimage" (158).[21] In a similar vein, the Marathi folklorist Durga Bhagwat (1974) has suggested that "every Maharashtrian" feels an "inner link" with the Varkari movement, "the kind of inner attachment one has for one's family and ancestors" (112).[22]

For all of these reasons, the Varkari Sampradaya has been a favorite signifier for Maharashtrian nationalist interests and the Maharashtrian Hindu Right for more than a century (Lele, 1987; 1995). Nativist and Hindu supremacist organizations like the Shiv Sena political party and the neo-brahminical Patit Pavan Sanghatana in Maharashtra have lost few opportunities to ally their image with the uniquely Maharashtrian "Hindu" spiritualism of the movement and its broad social base. But to view the early movement as an affirmation of Hindu social boundaries, or as a key moment in the evolution of the Marathi "nation," is a teleological reading of history that overlooks the intrusion of power in the re-presentation of the movement by various interest groups over the centuries. [23] Moreover, this perspective silences other very significant planes of identity and meaning that may have been important to Varkari participants at other times. If we put aside the nationalist narration of an emerging Maharashtrian consciousness, the Varkari Sampradaya appears less a Hinduist or pan-Maharashtra movement than a pan-agrarian and anti-statist movement. When viewed in this way, the contextual unity of the Varkari Sampradaya practitioners is not unlike the pan-agrarian identity proffered by the Shetkari Sanghatana.

One of the challenges to seeing the Varkari Sampradaya in these terms is that, because the movement's idiom of identity and action is spiritual, scholars (as well as Marathi nationalists) have typically pigeonholed it as a philosophical movement aimed at *religious*—rather than social—change (cf. Deleury 1960; Lorenzen 1987; Pande 1989; Sardar 1969). This classification distracts us from other very important aspects of the movement, and fails to take account of the fact that spiritual ideology in medieval Maharashtra was, to a very great extent, an ideology of social and political domination. This means that to contest the spiritual standards and to mobilize the masses in a display of solidarity was a collective threat to the existing order. For this reason, a few scholars have more recently begun to view the historical Varkari Sampradaya as a people's movement, with meanings and objectives deeply oriented toward social change (see especially Lele 1995; Omvedt and Patankar 2003). The issues at stake in this emerging debate are very much the same as those that concern us in attempting to understand the Shetkari Sanghatana: how to classify the movement; how to understand the influences on its ideology; and how to understand the intent and the subjectivity of its participants.

The Varkari Sampradaya movement emerged at a time when Maharashtra's brahmin community had begun to establish new forms of power and influence in the Marathi-speaking area. This began in the late twelfth century when, in exchange for protection and a continuing acknowledgement of brahmin supreme-caste status, the priestly community accorded moral legitimacy upon an ascendant dynasty exerting political rule from the non-Marathi south. In the arrangement, the brahmin lineages also expanded into new occupations of authority under the protection of their client regime. Brahmins came to dominate in roles such as moneylenders, tax collectors, state accountants, and royal bureaucrats, augmenting their already considerable spiritual authority with positions of more mundane social control (Pande 1989). Given the dramatic advances of brahmin power, it is probably not coincidental that, at almost the same time, a number of protestant bhakti movements emerged in the region. These movements rejected the moral justification of social hierarchy, called into question the role of the priesthood, and taught that everybody (regardless of caste, gender, or religion) could achieve direct personal communion with god without any need for priestly mediation.

One of these movements, the *Mahanubhava* sect, distanced itself from the priesthood by largely rejecting the power of the existing brahminically sanctioned Hindu pantheon. This sect posited a new and secretive conception of god that was without physical form (an imageless, or *nirguna*, deity) and thus unknown and unknowable to the brahmin establishment. Not surprisingly, brahminical authorities persecuted the Mahanubhavs and the sect eventually translated all of its texts into code to protect them from discovery and destruction (Feldhaus, 1988). The Varkari sect, on the other hand, was shown much greater tolerance. It became vastly more popular and openly proselytic across the Marathi-speaking land. Part of the reason for this greater tolerance and success was that the Varkari Sampradaya centered its devotion on an established regional deity named Vithoba, a folk god with a tangible image that could be visited in the temple at Pandharpur. More importantly, the Varkari saints publicly regarded this deity as an incarnation of the Puranic[24] deity Krishna. This meant that their sect, at least with regard to its patron deity, remained within the brahminically approved umbrella of established anthropomorphic *(saguna)* deities.[25]

That the Varkari's concept of divinity became established within the popular and dominant idiom of brahminically approved gods has led many scholars to view the emergence of the Varkari Sampradaya as little more than an elite brahmin effort to broaden their base of social support through limited spiritual reform "from above" (see Lorenzen 1987). Although this may account for some of the tolerance shown to the Varkari movement—and is not irrelevant to an understanding of the brahmin embrace of the Varkari Sampradaya today—what is lost in this is the way in which the Varkari participants used that popular idiom to interrogate and contest core meanings of brahminical ideology. Moreover, they also used it to practice new boundaries of collective identity that united Varkaris from a broad cross section of agrarian castes and communities.

We can see this in several patterns of Varkari philosophy and behavior. Where brahmin orthodoxy preached spiritual boundaries, caste boundaries, and untouchability, the medieval Varkari saints and practitioners who congregated and worshipped together defied those strictures by welcoming participants from any Hindu or non-Hindu social segments who were willing to oppose the ideological and social status quo. At a time when Dalits were required to live in isolated *wadas* outside the village walls and the orthodox punishment for shudras or Dalits who overheard the brahmins' Vedic[26] ritual incantations was to have their ears filled with molten lead, the Varkaris established a community of practice among men and women, shudras, Dalits, outcast brahmins (those excommunicated for violating orthodox brahmin norms), and even Muslims.[27] Where brahmins proscribed reading and writing (in any language) for people outside of the ruling castes or communities, the shudra, Dalit and Muslim Varkaris composed lengthy philosophical treatises and read these to assembled village masses. And where the Sanskrit-educated brahmin priests rejected the Marathi tongue of the masses as morally unsuitable for high ritual and philosophy, the bhakti practitioners in turn celebrated the vernacular and rejected brahminical Sanskrit as nothing more than a boundary-crafting strategy against the Marathi-speaking masses by a dominating Other. The Varkari challenge to the status quo was double edged. The Varkaris established a movement of noncooperation with the local authorities of state that not only rejected their brahmin symbolic capital and exclusive access to other tools of domination (such as reading and writing) but also— by eliminating brahmins as mediators between non-brahmins and the divine—sought to deprive them of their most established form of community participation and livelihood: the performance of, and payment for, ritual services.

Though the Varkari Sampradaya undoubtedly meant (and continues to mean) many different things to its participants, there are two further aspects that are enlightening as to its influence on community identity in the Marathi regions. One of these is the unusual duration and timing of the main annual pilgrimage. The walk, which can take up to two weeks depending on the distance of Pandharpur from one's home village, occurs in the lunar month of *Ashadh* (roughly late June through early July).[28] This timing, which frequently coincides with the onset of the monsoon and the busy season of planting and managing the *kharif* (wet season) crop, does not seem ideally suited to an extended absence from the fields. However, this timing falls at least seven months after the harvest of the previous year's kharif crop, and at least four months after the harvest of the dry season *rabi* (winter) crop, which in unirrigated areas is notoriously risky and subject to failure. It is important to recall that much of the Marathi land, particularly the rain-poor areas in which so many of the Varkari saints resided, was perpetually prone to famine—and famine would occur precisely in these interstices between the two major harvests. So, on the one hand, the Varkari march may have represented an opportunity to commune with god and earn divine merit that could help ensure a favorable monsoon. At the same time, it seems likely that the pilgrimage was, in many respects, a hunger march—carrying the rural poor across the countryside, where they were reverently fed by better-off families, in better-off villages and towns along the way.[29] The feeding of the pilgrims is a practice that continues today. This suggests, historically at least, a context of material opportunity (that is, being fed at a time of shortage) in the practice of the Varkari Sampradaya that may have been integral to the "spiritual" movement's capacity to elicit performances of identity that transcended barriers of caste and religion.

The reverence for and feeding of the pilgrims during the Varkari march highlights a second intriguing aspect of the pilgrimage: the Varkari Sampradaya also achieved a radical conceptual transference of the qualities of divinity and priestliness onto the peasantry and the qualities of sacred geography onto the countryside. Being ritually fed as an itinerant is more commonly reserved for brahmin priests and renunciate *sadhus* (ascetic holy men) than for otherwise ordinary peasants. Moreover, in most Indian pilgrimages, the focal point of the practice is the *destination*—what Mircea Eliade (1959) called the "sacred center." In the Varkari pilgrimage the focal point is the path, traversing the rural geographic periphery that lies *outside* the center of power (cf. Deleury, 1960; Engblom, 1987; Karve, 1988; Turner, 1974).[30] I have seen this ritual de-centering of the destination in my own participant-observations of the pilgrimage, in which arrival at Pandharpur seems more like a perfunctory denouement of the ecstatic journey than the apex of the experience. Throughout the course of the walk, the Varkari pilgrims consider themselves to be in spiritual communion not only with each other and with the spirits of the saints who trod the path before them, but also—because of the bhakti conception of direct inspiration—with divinity itself.

These unusual aspects of the pilgrimage led Victor Turner (1974) to view the Varkari pilgrim as a "total symbol" (208), by which he meant a single symbol capable of signifying all that the Varkari sect is and all that it reveres, including divinity. Thus, the practice embodies a subtle but revolutionary transference that, despite its negotiated accommodation of dominant ideology, was and is deeply threatening to power. In the pilgrimage, the profane geography of the nonruling castes becomes sacred space and the subaltern practitioners themselves become incarnations of the brahmins' own Puranic god. As I will argue later, there are important parallels between this Varkari ideology and the narratives about Bali and Bharat in the Shetkari Sanghatana.

The Varkari Sampradaya constructed a new oppositional boundary of community. It also constructed a new oppositional practice against authority. Unlike the Mahanubhav sect, it achieved these things through a popular idiom that was at its most central level sanctioned by power. Rather than regarding this as proof of ultimately elite origins of the movement, we need to keep in mind the ways in which the movement challenged existing contours of community, power, and place. Just as importantly, we also must acknowledge that, though in many ways an ideology of domination, the brahminical worldview of saguna deities was (and is still) a broadly influential and popularly accepted lens through which the masses perceived and practiced spiritual life. The core of that ideology constituted, in many ways, symbols that had (and continue to have) popular circulation and meaning across broad sections of rural Maharashtrian society. Hence, the popular culture and practices of the Marathi-speaking area presented the Varkari movement with both constraints and opportunities. The constraining aspect is that the movement leaders themselves were conditioned to experience and practice their lives within this web of meanings—and, even if they were not, outright rejection of the popular would have limited the movement's opportunities to expand among the masses that were. But this also presented opportunity. As a framework for a mobilizing discourse, it had tremendous potential for mass appeal and mass identification that, in the context of participation, could transcend other concurrent terrains of identity, and enable the Varkari sect to challenge brahminical rights to own and define those popular symbols.

The ideal king and the community of subjects
Another symbolic complex that is important for understanding contemporary dynamics of Maharashtrian collective selfhood is the rise and rule of king Shivaji's Maratha clan in the seventeenth century. Shivaji lived four hundred years after the first great saints of the Varkari movement. The founding of the Maratha Empire represents the first time that a major section of the Marathi-speaking land came under direct rule by an indigenous, Marathi-speaking community. Today, in the words of Maharashtrian political scientist V. M. Sirsikar (1999), the Shivaji era "has its permanent and persistent influence on the people of the region" and is "so deeply rooted in the Maharashtrian psyche" that "it is considered a sacrilege to make any adverse comments about the hallowed figure" (6). This historical period along with subsequent representations of it have been at the center of contests over varied constructions of identity and interest in Maharashtra since the days of Shivaji himself and certainly still are today. Examination of this era is important for our understanding of the Maharashtrian context and its influence on the Shetkari Sanghatana.

As noted earlier, in order to become a king Shivaji required legitimization from the brahmins who held the ultimate ideological power to sanction or reject local rulers. As a shudra, Shivaji would not even have been allowed to witness the Vedic rituals through which brahmins accorded approval of a king's rule. Shivaji found a solution: he arranged for some esteemed brahmins from Benaras to create a genealogy for his Bhosale clan that demonstrated an ancient descent from Rajput kshatriyas of northern India. This rendered Shivaji, now a kshatriya, fit to rule as a king—but, in turn, it also obliged him to accept the conventional kingly mantle of "protector of cows and brahmins" (O'Hanlon 1985). In the performance of this new kshatriya status, the Bhosale Marathas adopted outward signs of identity that were significantly more Sanskritized[31] than the other peasant castes that had previously been their more proximate social groups.

Shivaji, whose fledgling state was surrounded on all sides by powerful Muslim rulers from whom he had wrested large areas of the Marathi land, also needed popular support and recognition from the agrarian communities with which his Maratha clan was historically identified. To this effect, Shivaji made important concessions to the masses. He instituted land reforms, abolished many of the hereditary and oppressive jagir landlord privileges granted by previous rulers, introduced the *ryotwari* model of direct ownership by cultivators, reduced taxes, and effectively cut out much of the rural elite—including many local brahmin functionaries—from the administration of the state (Brahme and Upadhyaya 1979; Dhekane 1996; Wink 1986). For these reasons, his capture and rule of much of the Marathi land has sometimes been viewed as a mobilized agrarian response to the oppressive landlordism and taxation that had overburdened the peasantry (Rodrigues 1998).

Because of the ambiguities in his relationship to brahminical power *and* in his relationship to the agrarian masses, Shivaji and his reign constitute a pool of fluid signs with meanings that have been contested, embraced, or rejected by wide-ranging social groups. Modern historical narratives of Shivaji have regarded him variously as a Maharashtrian liberationist, a Hindu nationalist (purportedly opposed, on principle, to Muslim rule), a champion of brahmins, an enemy of landlords, and a champion of the oppressed shudras and Dalits of the agrarian castes (Gordon 1993; Jasper 2002; Laine 2003). Strikingly diverse ideological camps have used Shivaji narratives as a tool for invoking contours and histories of mass identity. These narratives were used by nineteenth and twentieth century advocates of Maharashtrian statehood to signify a deep history of ongoing struggle for Maharashtrian unification, and after statehood by Maratha elite elements of the Congress Party as a signification of a unified Maratha-kunbi block of interests. They were used by Maratha and brahmin interests during the Indian independence movement as a model of indigenous rule sanctioned by brahmin authority, and by shudra and Dalit interests in the Nonbrahmin Movement, as a model of low-caste achievement and low-caste rule. Starting in the latter decades of the twentieth century and continuing into the twenty-first, they have also been used by the Shiv Sena party (whose name means, literally, Shivaji's Army) to support narratives of Marathi and Hindu nationalism, and they have been proffered in the literature and rhetoric of the Shetkari Sanghatana, portraying Shivaji as a model king risen from the shetkari masses and a visionary of agrarian reform. Like the Varkari Sampradaya and its saints, Shivaji and the early days of the Maratha Empire have become so richly intertwined with ordinary cultural life that any movement attempting to craft mass contours of identity in the state must address the meaning of these symbols in one way or another.

In the early years of the Shetkari Sanghatana, movement leaders made overt efforts to claim and define Shivaji's meaning in public life. In speeches and lectures, Sharad Joshi often spoke of Shivaji's agrarian reform legacy as a contrast to modern agricultural policy. In 1988, he coauthored a book on Shivaji that depicted the king as an agrarian hero.[32] Although Sanghatana leaders continue to invoke Shivaji as part of their overall Marathi idiom, they have ceased relying on Shivaji as a core signifier. As one activist explained to me, Shivaji has become too deeply tied in the popular consciousness to the glorification of the state, the Maratha caste, brahmin legitimacy, and Hindu rule to be an effective symbol for an agrarian, antistate movement. But the Sanghatana's invocation of Bali Raj (as opposed to Shivaji Raj) as an alternative ideal of pro-agrarian rule cannot be completely understood in isolation from its contrast with Shivaji.

Today in Maharashtra, Shivaji's Maratha Empire has become widely associated with the Varkari Sampradaya pilgrimage. Shivaji, in the seventeenth century, was a contemporary of the last great Varkari saint, a kunbi from the Desh by the name of Tukaram. Although historians doubt whether the two had actually met, and although no known documents from the seventeenth century clearly associate Shivaji with the saint (Laine, 2003), the association has become deeply embedded. Traveling through Maharashtra today, one sees an abundance of imaginative "historic" portraits of Tukaram and Shivaji together—the gentle kunbi saint extending his hands in a blessing to the Maratha nationalist icon. The effect of this symbolic merger of Shivaji and the Varkari saints has been to further ambiguate and universalize the boundaries of identity signified by both sets of signs.

Efforts to blur and redraw social boundaries through the linkage of these popular symbols have a long history. Though widely evident today in a broad array of cultural productions, this intention can be traced to the eighteenth century rise of the brahmin peshwas in Maharashtra. The peshwas, the Konkani brahmin prime ministers of the Maratha Empire, acquired direct control over the empire in 1749. Although they reduced the Maratha descendants of Shivaji to mere figureheads and presided over unprecedented expansion of territory and influence for the empire, the Shivaji narrative was never replaced with a prominent narrative of peshwa rule as symbol of Maharashtrian identity (Laine, 2003). Quite the opposite of replacing the Shivaji narrative, the peshwas, who had a deep interest in maintaining a popular impression of Maratha rule, promoted it. One reason for this was that anti-brahmin sentiment continued to be a significant force among the masses (Lele 1995; Omvedt 1976). A second reason was that the Maratha narrative of an emerging Marathi state had already become firmly established as a popular legitimization of the empire and as a claim on the Marathi-speaking lands that were still held by other powers (Laine 2003). But probably the more important reason for this was that a façade of Maratha rule could more easily signify "local" rule to the preponderant Maratha and kunbi communities of the empire. The Chitpawan brahmin peshwas were a small minority who hailed only from the Konkan region and could command no broad panregional identification from the Marathi masses.

For many of the same reasons, the peshwas also had a vested interest in linking the heroic Shivaji narrative to the Varkari movement of the masses—in effect, attempting to re-signify the Varkaris' rural mobilization and anti-brahminism as an authorization of the Maratha rule that the peshwas were purportedly working to preserve. Toward this end, the peshwas commissioned new biographical accounts of Shivaji and hagiographical accounts of the saints that linked these narratives and further cloaked the real brahminical underpinnings of imperial authority (ibid.). This brahminical move toward resignification of the Varkari saints probably underlies many scholarly readings of the Varkari Sampradaya as "reform from above" rather than reform by the masses—but it is a move that did not occur on any significant scale until five hundred years after the founding of the movement.

Nonetheless, these two pools of signification—Shivaji and the Varkari Sampradaya—for the last two hundred years have been firmly fixed as the backbone for narrations of Marathi unity and other constructions of mass identity within the state. Sometimes these narrations are competitive, sometimes overlapping, and sometimes oppositional, but they are crucial to any understanding of the contours of Maharashtrian identities today.

From Nonbrahminism to ruralism
When the peshwas helped create symbolic links between Shivaji

and the Varkari movement, they had yet another solid reason for doing so. Though the Maratha royal families had been reduced to mere figureheads of the empire, important Maratha households continued to exert significant power in social and cultural affairs. More importantly, the Maratha elite had become antagonistic to the resurgent brahmin authority—both political and cultural—with which they had two centuries earlier reached a political and ideological compromise. This formal brahmin authority continued another two centuries until well after the formal establishment of British rule, with the highly educated Chitpawan brahmins retaining many influential posts and positions in British colonial state officialdom (Lele 1995; Omvedt 1976).

The most important popular expression of broad caste ruralism to emerge in the last two centuries was the Nonbrahmin Movement, which was specifically focused on the Otherness of this resurgent brahmin authority. The Nonbrahmin Movement, largely associated with the Satyashodhak Samaj (truth-seeking society) led by Jyotirao Govind Phule (1827–1890), rejected brahmin cultural, political, and social dominance. It advocated broad social and ideological reform to improve the lives of the middle and lower agrarian castes. This movement, which continues to inform Dalit and other lower-caste mobilizing in the state (as well as the Shetkari Sanghatana), expanded dramatically in the latter nineteenth century. It attracted adherents from towns as well as villages, and had the feel of revolution. In villages across the state, laborers refused to work in the fields of brahmin landlords unless they improved the working conditions and relinquished their claims to superiority. Tenants left their land fallow in protest of the landlords' exorbitant rents (O'Hanlon 1985). Across the Marathi regions, peasants ousted local brahmins from their ancestral villages and they grabbed the brahmin land for their own use (Brahme and Upadhyaya 1979). Hence, in this respect, the Nonbrahmin Movement was about far more than cultural ideology. It also sought a redistribution of agrarian resources, directed against the most proximate and prominent wielders of expropriative power and wealth.

The Nonbrahmin Movement was embraced and even financially sponsored by elite Marathas. In individual villages, it was often the Maratha patil who encouraged Nonbrahmin activism (O'Hanlon 1985). In the towns, the movement was financially sponsored by the elite Maratha houses of Baroda and Kolhapur (areas where royal Marathas continued to reign, if not exactly rule), who were the immediate contenders for formal and ideological power in the absence of brahmin authority.[33] The Nonbrahmin Movement thus functioned as a tool for uniting the masses, but the movement was highly advantageous to specifically Maratha political interests. The Marathas and kunbis together as a broad, affiliated caste "cluster" (later to become a powerful voting constituency) can be traced to this period (Attwood 1988).

Despite this, most of the Nonbrahmin ideologues and activists were neither elite Marathas nor primarily concerned with Maratha interests. Phule, for example (who was from the Mali caste), occasionally used the term Maratha as a broad reference to all of the communities of Maharashtra that are not brahmin—but because his usage of the term included Dalits, adivasis, and non-Hindus it effectively held no specific "caste" distinction and reserved no pride of place for the Marathas themselves (T. L. Joshi 1996). Moreover, although Phule participated in the seemingly mandatory practice of narrating the life and meaning of king Shivaji, he used the Maratha king to signify an ideal of agrarian shudra and *ati-shudra* (Dalit) greatness, effectively rejecting the superior kshatriya claim of elite Marathas (Keer 1974; O'Hanlon 1985).[34]

Phule's impact on Maharashtrian culture, society, and politics has been enormous (cf. O'Hanlon 1985; Omvedt 1976; Zelliot 1982). A tireless activist and prolific writer, many of Phule's contributions are deeply reminiscent of the Varkari Sampradaya: he developed an elaborate critique of brahminically-endorsed social structure, fought for the equality of all castes and communities, advocated the elimination of priests from marriages and other rituals, and even articulated (and observed) a new form of religious practice.

Though he was an urban intellectual living in the city of Pune, the majority of Phule's Nonbrahmin critiques were focused on the welfare of the agrarian masses. In his collection of Marathi poems called *Brahmanace Kasab (The Cleverness of Brahmins)*, Phule attacked brahmin social and financial institutions in the villages. He described the gullibility of agriculturalists who went into debt (most often to brahmin moneylenders) in order to pay brahmins to preside over brahminically mandated rituals, all the while remaining blissfully unaware of their exploitation. By the time of the so-called Deccan Riots in Western Maharashtra in 1875, Phule had already become perhaps the most prominent contemporary advocate of agrarian change in Maharashtra (and India in general). His lobbying efforts through the Samaj were influential in securing provisions for tax reform, levy reductions, and debt relief in the Deccan Agriculturalists Relief Act of 1879 (T. L. Joshi 1996).

By the 1880s Phule had become almost exclusively concerned with the plight of the rural areas. Delivering lectures in villages and towns, he expounded on the ways that noncultivating villagers and town-dwellers alike were complicit in exploiting the predominantly low-caste food producers of the village, and he argued for the necessity of agrarian unity. Although Phule's criticism initially targeted the brahmin castes, he eventually broadened his notion of the oppressive Other to include a wide range of town-dwellers and nonagriculturalists, regardless of their caste. Phule established a foundation for rural mobilization in which being an agriculturalist was more important than any other identity or affiliation. By the time he compiled his lectures of 1882–1883 into one of his most famous books, called *Shetkaryaca Asud (The Shetkari's Whip)*, his vision of a united rural community had evolved into what today sounds very similar to the Shetkari Sanghatana's conceptualization of Bharat and the community of shetkaris.

Phule never ceased to view the ideology of brahminism as a hindrance to social enlightenment and rural unity. He was adamant that caste-ism and social hierarchy were brahmin inventions maintained through brahminical ideology, and he set out to reconstruct a debrahminized spiritual subjectivity that would free the minds of the agrarian castes. Central to his reconstructed spiritual world was the veneration of the demon king Bali, which he saw as an alternative not only to the Puranic gods and the Hindu nationalist ideal of *Rama Raj*[35] but also to the Marathas' and brahmins' nationalistic icon of king Shivaji.

Phule's use of Bali has been a significant influence on Sharad Joshi and other leaders in the Shetkari Sanghatana—but Phule was certainly not the first Maharashtrian to depict Bali as a heroic ideal for the agrarian masses. As I will explore in more depth in later chapters, Bali is a well-known character in pan-Indian Hindu mythology. He is a multivocal symbol, frequently represented in the ancient Puranic literature with a mixture of both admiration and contempt. Folktales and folk rituals associated with Bali were not at all unfamiliar to the Maharashtrian agrarian castes before Phule. In fact, Bali has long held a distinctive significance in rural Maharashtra that is unparalleled in any other part of India and this significance almost certainly had anti-brahmin cultural overtones long before Phule. Phule's contribution was to greatly expand the openly politicized use of the demon king in popular discourse and writing, extensively elaborating on Bali and Bali's mythological rule as a metaphor for a pre-Aryan (in other words, pre-brahmin) golden age of equality. In a brilliant but often fanciful reconstruction of history, he interpreted each of the five major incarnations of the brahminical god Vishnu as a collective historical memory of consecutive stages of Aryan conquest. In this telling of history, the fifth avatar, the dwarf brahmin named Vamana who unseated Bali and deposed him to the nether world of Patal, represented the final and most crushing stage.

But Phule also recognized the importance of the popular ritual calendar of "great tradition" Hinduism, which is dominated by Aryan, brahminical rituals and symbols. Although a rationalist, he did not attempt to dissuade the agrarian castes from observing these rituals. Instead, he used these as a platform for advancing his Satyashodhak ideology within the key agrarian festivals and rituals, excavating each festival and ritual for elements of a pre-brahminical egalitarian ethos. For example, he attempted to resignify important practices in the major annual Hindu festivals of *Navaratri, Dasara,* and *Diwali* as ritualized reenactments of the calamitous confrontation between Bali and Vamana—much in the way that some of these are described and performed today by many Shetkari Sanghatana participants. In the words of historian and Phule scholar Rosalind O'Hanlon, Phule constructed "a complete rival interpretation of the conventional Hindu religious year...[such that] adoption by a potential follower would not have entailed any sense of the dislocation that would have accompanied the complete destruction of traditional categories of thought and practice" (1985, 160).

Many of the ideas in his "rival interpretation" were novel twists that are difficult to defend with scientific historical evidence. But Phule's notions had cultural resonance. His ideas were influenced by historically deep, heterogeneous, lower-caste interpretations of popular Hindu practices in Maharashtra. The extent of his influence in communicating resistant meaning through Bali—much of which may have already existed as less vocal "hidden transcripts" in popular culture (Scott 1976, 1985)—is difficult to determine. What is clear is that much of the significance of Bali described in Phule's writing continues to be discernable today in the expressions and practices of rural Maharashtrians.

Toward independence and back
In the late nineteenth and twentieth centuries, ideologies of mobilization in Maharashtra increasingly hinged on narratives of mass unity, spanning multiple castes or classes. These were in direct competition with narratives that focused on social boundaries and divisions, and often played out as battles for definitional control of key signifiers representing Hindu "tradition," the Marathi language and Maharashtrian-ness, the

Varkari Sampradaya, and Shivaji's Maratha Empire.

Competing narratives struck different chords in different communities and regions of the state. Various camps invoked popular symbols of Hinduism to represent the ideology of an upper-caste oppressor, the ethical imperative of a self-determined Hindu nation, or the unified opposition of colonized subjects against British rule. In the Nizam's Marathwada, where Marathi-language public schools and Marathi newspapers did not exist until well into the twentieth century, the Marathi language was symbolically represented by some of the masses as the voice of the oppressed (Stree Shakti Sanghatana 1985). In other regions where Marathi was relatively well established in formal institutions, the language was widely celebrated as the national tongue of the once glorious Maratha kingdom and of the future state of Maharashtra. Among many agrarian movements, the Varkari Sampradaya represented a legacy of agrarian caste unity as well as a challenge to monolithic constructions of Hindu identity, but for Marathi nationalist groups it also symbolized a precursor of Maharashtrian collective will. Shivaji increasingly appeared at the representational fore of contested narratives on the ideal of Hindu self-rule (for the brahmin-dominated national independence movement), the ideal of local Maharashtrian rule (for Maratha political aspirants), or the ideal of social advancement (for the kunbi and Dalit masses).

Shortly after Indian independence in 1947, the narratives of Maharashtra statehood and corporate Marathi identity returned to center stage in the *Samyukta* (unified) Maharashtra movement that ultimately resulted in the union of the four Maharashtrian regions into one state. As I have indicated briefly previously, the Samyukta movement was not universally embraced in the Marathi-speaking land. Both in Marathwada and Vidarbha, the notion of a singular Marathi identity was (and still is) contested by calls for the independent statehood of these regions. Critics and skeptics of a unified Maharashtra included some key figures within the Maharashtra Congress Party, such as the Dalit leader Dr. B. R. Ambedkar, who was concerned that Marathwada's and

Vidarbha's large Dalit constituencies would be underrepresented politically in a Maratha-brahmin dominated unified state. The struggle for (and against) Maharashtra statehood heightened the cultural debate on different Maharashtrian identities and the fields of opportunity afforded by competing constructions of collective selfhood. By the time of Maharashtrian statehood in 1960 another clear contour of disunity was also emerging: the division between the new post-independence industrial-political elite and the rural masses.

This contour of opposed groups, not wholly unlike the one articulated by the Varkari saints seven hundred years earlier, began to reach new prominence with the imposition of centralized planning. Under Jawaharlal Nehru's twin imperatives of import substitution and rapid industrial development, planning was oriented toward using primary resources and agricultural goods to fuel the growth of the urban domestic economy. The resulting rift in the pace of economic and social development between the cities and the villages was not unique to Maharashtra. In the 1970s, the rift lifted the charismatic youth leader Jayaprakash Narayan to national prominence and made him an inspiration to an entire generation of left-leaning leadership, as he sought to reinvigorate the Gandhian spirit of rural service. Narayan called on urban activists and disillusioned students throughout the country to go to the aid of the villages that had been left behind.

In Maharashtra, the urban-rural rift was prominent. Amidst the rapid growth and rising economic vitality of urban centers like Pune and Mumbai, most of the Maharashtrian hinterland continued to suffer from poor soil, insufficient rain, undeveloped irrigation, and a level of agricultural productivity well below the national average. By the early 1980s, when Sharad Joshi and the Shetkari Sanghatana began decrying the internal colonization of Bharat by a new class of urban-industrial "postcolonial" oppressors, the salience of the claim and the opportunities afforded by a collective "shetkari" identity must have seemed self-evident to many. Just as importantly, it must have had a familiar ring. When the Sanghatana staged its first major statewide propaganda campaign in 1983 as a procession of Shetkari Sanghatana "pilgrims" in the annual Varkari march to Pandharpur, it could not have more clearly signified that the black British, regardless of their caste, were the new ideologists and oppressors for the rural masses to resist. And when it called for a restoration of Bali's rule in Bharat, the symbolism of Bali as a truly pro-agarian king, unjustly deposed by the spiteful self-interest of an empowered minority, could not have more vividly pronounced the Sanghatana's claim to the mantle of rural unity and justice.

Conclusions

Maharashtra is a large and young state with varying regional histories, social and economic patterns, and natural environments. Nonetheless, the state does exhibit prominent points of historical, social, and cultural continuity across the regions. These points of continuity, in addition to serving as the underlying justification for the union of the regions into the single ethnic-linguistic state of Maharashtra in 1960, also constitute key pools of signification through which other non-nationalist constructions of identity and oppositional unity have been narrated.

With few exceptions, most agricultural areas of the state are relatively undeveloped and relatively underproductive in comparison with national standards. Despite this relatively weak position of the agrarian economy, market participation by Maharashtrian agriculturalists has become unusually high and is (compared with other states) well distributed across agrarian castes, communities, and landholding sizes. This high level of production for the market, though of greater or lesser importance in the overall economic strategy of differently situated landholders, represents, at least in some contexts of opportunity, a potential point of continuity in agrarian interests that can transcend other Maharashtrian contours of identity and interest. This potential for agrarian unity is also supported by the fact that, unlike the agricultural sector, economic growth and state-sponsored development in Maharashtra's industrial sector since independence have become the highest in the country. This has encouraged the perception of a social rift between urban-industrial elites and rural inhabitants (sometimes articulated as a rift between the more developed and less developed regions) that is reminiscent of earlier expressed antagonisms between local agents of the state or beneficiaries of power and the agrarian castes. These antagonisms have historically been played out, to a large extent, as struggles to control the production of meaning in the broadly shared symbolic pools that have been used to signify varied constructions of collective selfhood and unity in Maharashtra.

Despite these evident struggles over meaning, it has often been noted that a distinctive aspect of Maharashtrian history, prior to the rise of the Shetkari Sanghatana in the 1980s, is its "absence of any independent and effective organized movement of the rural poor" (Upadhyaya 1980, 213). It is true that—aside from the ongoing adivasi (tribal) movements of certain districts, and the occasional uprisings in the nineteenth century Deccan and independence-era Telangana—organized agrarian revolt in Maharashtra has been far less prominent than in some other states. At the same time, it is a mistake to suggest that the poor have not responded to the ideological and material mechanisms of their oppression.

Part of the challenge in seeing a history of collective agency of the masses lies in the fact that Maharashtra's agrarian communities have more typically organized not as individual classes but in broad social groups in common opposition to a source of systemic and mutual oppression. Moreover, they have commonly done so through an idiom of ideas and action that is partially accommodating of the dominant ideas and system. We can see this in the Varkari pilgrims' rejection of hierarchy and privilege through idioms and practices largely endorsed by power; in the agrarian efforts to reclaim Shivaji as a shudra king; in the Nonbrahmin attempts to endow existing Hindu practices with critical and oppositional meaning; in the vacillations between Marathi and regional identity in Marathwada and Vidarbha; and in the strategic maneuvers of different agrarian castes and communities to resignify their ideological status. All of these can be seen, in their own way, as efforts of the rural poor to reconfigure the social and political opportunities available to them and to maximize their collective capacity to struggle for social and economic change. These strategic efforts occur within a larger interpretive community that is constrained and shaped by "the language and ideas of the state," but they are also dialogically engaged with these dominant ideas and their forms of signification.

As I will show in the coming chapter, the Shetkari Sanghatana is also a dialogical formation within the larger interpretive community of Maharashtra. While it distinguishes itself from other causes and other movements within the state, each plank of its signification is inevitably tied, in varied degrees of accommodation and contestation, to the larger field of interests and significations of identity that characterize contemporary Maharashtra.

Notes

[1] That is, "of Pune."

[2] The official language in Goa is Konkani. There is a highly politicized debate as to whether Konkani is a separate language or a coastal dialect of Marathi.

[3] The foregoing history is compiled primarily from Dikshit, 1985; Karve, 1968; and Mansingh, 1998.

[4] In July 1998, during my fieldwork, this number of districts rose to a total of thirty-two as two new districts called Washim and Nandurbar were carved out of the existing districts of Dhule and Akola, respectively (*IE* 7/1/98, 2). Among the arguments offered for this change were distinct historic differences and economic challenges within the proposed new district boundaries that necessitated independent administrative attention. While this points to ongoing disputes over spatio-historic identities in the state, the most probable reason for officially creating the new districts would have been to gerrymander advantageous constituencies for the then ruling Shiv Sena state government.

[5] Nitin Gadkari, Maharashtra Public Works Minister (quoted in *IE* 12/20/98, 2).

[6] Sirsikar, citing the 1981 census of India, indicates that the Konkan's Ratnagiri district alone accounts for 45 percent of all Maharashtrian labor migration into Mumbai. Although this nexus between rural households and urban labor opportunities is particularly strong in the Konkan, it is also very significant in many areas of the plateau region (see especially Dandekar 1999).

[7] I have heard Patal translated for me in this way many times by informants, but also by Maharashtrian folklorist Dr. Sorojni Babar (personal conversation). Pandit Mahadevshastri Joshi, in his *Bharatiya Sanskrutikosh (Encyclopedia of Indian Culture* 1962), gives one definition of *Patal* as "the coastal edge of Maharashtra." Molesworth (1996, first published 1831), in his still unrivalled dictionary of the Marathi language, indicates that the word derives from roots for "subterranean" or "hidden."

[8] An interesting example of this is the case of the Konkanastha Chitpawan brahmin community. The origin myth that I have heard many times distinguishes their lineage from other Maharashtrian brahmins by attributing their patrimony to a group of shipwrecked sailors who washed up dead on the ocean shore and were brought back to life.

Interestingly, the small (and now mostly relocated) Ben Israeli Jewish community of the Konkan also attributes its origin to a group of shipwrecked sailors who one day washed up on the shore—thus distinguishing themselves from the Baghdadi Jews of Pune city, in Western Maharashtra, who trace their own appearance in India to an overland migration from Iraq. I have heard these origin stories from members of the Baghdadi and Chitpawan communities in Pune. They are also mentioned in Chopra (1998).

[9] Formerly and still frequently written in English as "Poona."

[10] Scholars disagree whether Shivaji actually considered the brahmin Ramdas his guru, and it seems possible that this is a largely imaginary reconstruction by brahmin biographers of Shivaji and hagiographers of Ramdas. Nonetheless, the relationship is today popularly accepted (see Laine 2003).

[11] Under the Nizam, these two very different systems had a similar impact on the peasantry. The *jagirs*, lands to which the Nizam had granted independent revenue title as rewards or guarantees of loyalty, were essentially independent fiefdoms, free to tax and administer according to the *jagirdar's* whims. These constituted about 40 percent of Telangana. The remaining 60 percent of Telangana was mostly what was called *diwan* land, in which the Nizam held sole revenue title. Land tenure in diwan areas was theoretically based on an owner-cultivator system with revenue paid to the state. These areas were contracted out to revenue agents, under the title of *talukdars*, who collected taxes on behalf of the state. In practice, the talukdars exercised nearly total control of the land, extracting additional revenue for their own purses and often becoming the virtual owners of villages under their jurisdiction. Although tenure in the diwan lands was restored to owner-cultivators in a British inspired reform during the latter nineteenth century, in practice this only led to the reproduction of the talukdari system on a smaller scale, in which peasant owners converted themselves into landlords and, in collusion with a rising class of moneylenders, rented or mortgaged land to poorer tenants (see Brahme and Upadhyaya 1979).

[12] Maharashtra Public Works Minister Nitin Gadkari cites the number of villages in Maharashtra as 49,522 (quoted in *IE* 12/20/98, 2).

[13] This boundary of "Hindu" is difficult to define, even for the simple expedience of collecting census data. The definition employed in the census is based on the Indian constitution's own "other than all other

religions" test—that is, anyone who is not Muslim, Buddhist, Christian, Sikh, Jain, Zoroastrian, or Jewish.

[14] As these and other scholars have argued, Hinduism as a category is in many ways an invention of Indologists, missionaries, and colonial authorities who sought to classify and apprehend Indian heterodox religious practice in relation to scripture-based faiths of Judaism, Christianity, and Islam. Oberoi extends this more broadly, to include the reification of South Asian Islam, Sikhism, and other constellations of ideas and practices that he argues were historically fluid and resistant to classification: "Ironically, the post-colonial state has continued employing the census strategies of the colonial state. But even after the question 'What is your religion?' has been repeatedly raised now for several generations, there are people in the country who are unable to conceptualize their beliefs and rites in terms of a monolithic universal community" (1994, 10).

[15] The dominant caste concept articulated by Srinivas identifies a caste as dominant when it 1) preponderates numerically over society, *and* 2) wields preponderant economic and political power. Srinivas also points out that a caste group can more easily become dominant if its position in the caste hierarchy is not too low. This consideration acknowledges that dominance must, in one way or another, accommodate the (brahminical) ideology of hierarchy (Srinivas 1996).

[16] *Varna* is one of the two conventions of social organization in South Asia that are normally subsumed by the less nuanced English term "caste." Varna refers to the division of society into four hereditary categories, roughly equivalent to the social function of each group: *brahmin* (priests and intellectuals), *kshatriya* (warriors and kings), *vaishya* (a broad spectrum of middle occupations), and *shudra* (peasants and workers). In brahminical ideology—and hence, also to a very large extent in practice—these four groups are considered to be hierarchically ranked, with brahmins at the apex. The other convention—*jati*—refers to distinct communities that are defined sometimes by occupation or religious practices, and most frequently represent an endogamous kin group. Most, but not all, jatis are regarded as falling within one of the four varnas, but competing social groups may disagree on precisely which of the varnas a jati belongs to. Moreover, in practice jatis are also ranked in relation to each other even within the same varna. This relative status of different jatis is also often a point of contention.

[17] The practice of granting jagir rights to loyal families was common under the Maratha regime, as it was in the Nizam's Hyderabad. Scholars disagree on whether the jagirdari system under the Marathas was as socially oppressive as it was in Telangana, but it does seem that Shivaji was quick to institute a number of agrarian reforms early in his career—including some reduction of the jagirdari system and the establishment of the *ryotwari* system (cultivator-ownership of land and direct payment of revenue to the king) in the non-jagir lands (Brahme and Upadhyaya 1979). Interestingly, this aspect of "relieving the peasant burden" is frequently invoked in popular Shivaji narratives, whether Hindu nationalist or the oppositional secularist (see Laine 2003. Also Sharad Joshi et al 1988).

[18] For an excellent discussion on contexts of speech and variable caste terms of reference in the village of Apshinge, of Satara District, in Western Maharashtra, see Schlesinger (1988).

[19] It is noteworthy here that the Congress Party, long considered the bastion of Maratha power in Maharashtra, had at the time of my field research lost much of its electoral dominance across the state. The exception to this was the predominantly sugar-producing districts in the southern half of Western Maharashtra. In the 1996 general elections, Western Maharashtra was the only region of the state where the Congress Party achieved a majority, taking a total of six districts (*IE* 12/15/97).

[20] The theological origins of bhakti far predate the twelfth century, and are normally traced to the philosophy of the Bhagavad Gita composed sometime around 200 BC. Whereas the earlier Vedic texts emphasized the role of brahmins as mediators between individuals and the divine, the Gita suggested that anyone could develop a personal relationship with divinity through intense devotional worship.

[21] Quoted from Karve's classic account "'On the Road': A Maharashtrian Pilgrimage," reprinted in part in Zelliot and Berntsen (1988). The essay was originally published in English in the *Journal of Asian Studies* 22:1 (1962), 13–29.

[22] Quoted from Philip C. Engblom's Introduction to *Mokashi*, 1987.

[23] This also overlooks a wide range of other deeply established pilgrimage practices that either transcend Marathi boundaries or are confined to other cultural geographies (such as regional river valleys) within the modern Marathi state—any of which we could today look back at as contributors to other geographic constructions had the political cartography of the region evolved differently.

[24] The Puranas are a collection of works of uncertain authorship and uncertain antiquity that were probably composed over a period between the fifth century BC and the fourth century AD. They are a collection of expositions on theogony; cosmology; the genealogy of great kings and sages; and religious, social, and political practices. They are also the accepted principal source of ancient knowledge on the pantheon of Hindu gods and goddesses. Central to the Puranic pantheon is the trinity of three gods: Brahma (the creator), Vishnu (the preserver), and Shiva (the destroyer). In the Puranas, Krishna is held to be an avatar (incarnation) of Vishnu.

[25] The idea here of *public* regard for Vithoba as an incarnation of Krishna is important. Some recent readings of the Varkari saints' poems suggest that, on a deeper philosophical level, the saints may have more skeptically regarded all deities as products of human imagination and merely symbolic of a higher ideal. See, for example, Omvedt and Patankar (2003).

[26] The Vedas, sacred texts that predate the Puranas are (along with the great epics) the other key source of ancient knowledge about Hindu cosmology and ritual. Unlike the popular lore of the Puranas, however, much of Vedic content consists of hymns and incantations to be used in rituals and sacrifices, and many of these were considered inappropriate for the ears of shudras.

[27] Bhakti, as a philosophy of direct, inspirational communion with the divine, shares much with the mystic sufi philosophy of Islam. While bhakti worship in India seems to have deeper roots, the greatest flowering of collective bhakti movements occurred during a time of expanding contacts between India and the Islamic civilizations to its west—most notably, in the late twelfth century, with the arrival in the subcontinent of Muslim missionaries from the sufi sects of Arabia and Persia. Similar to the bhakti sects, the sufists practiced a personal form of devotion that contested of the scriptural orthodoxy of their own Islamic clerical establishment. And the sufi teachers—just like those of bhakti—used devotional poetry, music, and dance as favored mediums for meditation, worship, and discourse. By the thirteenth century, with the formal establishment of three orders of sufism in India and a rising tide of organized bhakti sects among the Hindu majority, it was not uncommon to find Muslim sufists and Hindu bhaktas (bhakti practitioners) mixing together in their common quest for spiritual enlightenment.

[28] Some Varkaris make the pilgrimage twice per year, and a very small number make the journey every month. However, the primary pilgrimage that is important to all Varkaris occurs in Ashadh.

[29] Maharashtrian archeologist D. D. Kosambi (1994 [1962]) suggests that pilgrimages to Pandharpur—as well as the seasonal timing and some of the actual routes traveled—may trace back to the Mesolithic period. Kosambi argues that the timing and routes of movement seen in the Varkari sect make little sense for an agricultural society, but would have made a great deal of sense for early pastoralists concerned with seasonal rains and pasture. This assessment is not without complications. As anthropologist Anatoly Khazanov points out, there is little evidence that pastoralism even existed as a mode of resource utilization during the Mesolithic (personal communication). Even if Kosambi were correct, however, this does not negate the idea that the popular emergence of the Varkari Sampradya as a pilgrimage cult of agriculturalists begs a consideration of its material logic.

[30] Some other major pilgrimages in India—notably those to Kedarnath or Badrinath in the Himalayas—do consider the journey an important means of acquiring merit through the personal *sacrifice* of embarking on a challenging and rigorous trip. However, as Marathi culture scholar Philip Engblom (1987) points out, this is quite different from the Varkaris' treatment of the path itself as a moveable destination and the entire journey as a point of communion.

[31] Sanskritization, a term offered by M. N. Srinivas (1987, 1996) refers to a caste's or community's adoption of social and cultural practices that are imitative of a hierarchically superior caste or community. Srinivas views this largely as a conscious or semiconscious strategy for upward mobility, enabling greater association with superior groups and recognition from all groups of a more advanced social standing.

[32] *Shetkaryanca Raja Shivaji* (Shivaji, King of Shetkaris), 1988. In Marathi.

[33] Thanks to Lee Schlesinger for drawing my attention to the contradiction between the nonbrahmin movement as an agrarian movement and as one simultaneously supported by "town centric" Maratha ruling houses.

[34] It is also noteworthy that Phule rejected both Maratha and brahmin aspirations for political independence from Britain. Phule felt that, on balance, the British presence would help to facilitate a reformation of Indian society through the moderating influence of western civil law, and what he saw as Britain's democratic and egalitarian ideals. He and many of his fellow Satyashodhak Samajists took this very seriously; one of the

initiation rituals for new members of the Samaj was an oath of allegiance to the British crown.

[35] Rama, the hero of the epic Ramayana, is widely considered to be an incarnation of Vishnu and (by many Hindus, and particularly Hindu nationalists) as an exemplar of the divine king and ethical rule. Considering Phule's pro-British sentiments and anticommunalist ideas, it is not surprising that he was opposed to ethnic and religious nationalism. Phule did believe in (eventual) self-determination for India, but conceptualized the ideal state in very different terms. In the words of biographer T. L. Joshi (1996), Phule's idea of the ethical state was "the community of a region, based on freedom and equality" (47). Thus he considered the precise contours of a state "region" to be less important than the state's ability to uphold democratic, secular, and egalitarian values.

3—The field of signs

While propagating the thought of the Shetkari Sanghatana, activists should abide by certain rules. Though the movement is being built on a larger scale than ever before, activists should not deride earlier movements and agitations that have preceded us. It should be remembered that the earlier movements helped to open the eyes of the shetkari.

—Sharad Joshi,
Shetkari Sanghatana Vicar ani Karyapadhdati
(Thought and Practice of the Shetkari Sanghatana)

If there is one site that is most representative of the central leadership and ideology of the Shetkari Sanghatana, it is Angarmala. This is the headquarters from which Sharad Joshi and his associates run the core activities of the movement. Angarmala is located about 25 miles (40 kilometers) from the city of Pune, in the Desh subregion of Western Maharashtra, the historic heart of the Marathi homeland. Leaving Pune, the landscape slowly changes from one of deeply gnarled traffic, suburban housing developments, and industrial complexes, to an airy green and semiprosperous agricultural zone. It is surrounded by tall hills, steeped in the legends of the seventeenth-century king Shivaji and of the Varkari saints Tukaram and Dnyaneshwar.

This headquarters, just beyond a small, sheet metal Shetkari Sanghatana signpost on a country road, is located on Joshi's own land, where he settled and became a gentleman cultivator (and, soon thereafter, an agrarian leader) in the mid 1970s. The land around Angarmala is speckled with Joshi's assorted experiments with crops, land improvement, and water management. The compound itself, a short walk along a stone path from Joshi's house, is built for utility rather than showiness. The walls are rough concrete, the floors are stone, and the bare rafters are topped with corrugated metal roofing. In most respects, it is an unassuming space. On an ordinary, quiet day only the name "Angarmala" painted on one of its exterior walls would suggest that this is an important center of discontent and social mobilization. The name, alluding to a Maharashtrian history in which rural folk have been simultaneously lionized and exploited, means "garland of burning coals."

Although Angarmala's interior is as utilitarian and simple as it appears from the outside, it is clear that this is a hub of activity. It is part dormitory for visiting activists and part assembly hall for training camps, meetings, and other gatherings, where all attendees sit on ordinary *dhurrie* rugs spread on the floor. There are steel cots, blackboards, a small kitchen, a half dozen rough-hewn stone tables for serving meals, and a Ping-Pong table for diversion during long gatherings and work sessions. There are also two anterooms that serve as additional dormitory space, archives, and the offices of the Sanghatana's secretary, a man who would become an important informant and a welcome companion during my many stays at Angarmala. Scattered about the main rooms are assorted propaganda posters, photographs of famous Sanghatana rallies, memorabilia and accreted images that are shared touch points within the movement. On one wall is a large painted image of the movement's logo—white lettering on a disk of searing red declares "Shetkari Sanghatana" in Marathi. This logo is the centerpiece on the Shetkari Sanghatana flag, also depicted throughout the room, on which the logo hovers like a blazing sun over a horizontal field of green.

In this chapter, we will examine the Shetkari Sanghatana through the macro-level lens of its central leadership and ideology. In the first part of the chapter, I will consider the Sanghatana's most loudly articulated demand, the One Point Plan for remunerative prices, and assess the credibility of leaders' claim that this is a unifying issue for rural Maharashtrians. In the second part of the chapter, I will look closely at the wide assortment of signs, strategies, and resources deployed within the movement that enhance the credibility of this claim and help create ideas and experiences of common interest in the movement.

"Price" and economic rationality in the experiential field

Despite the great fluidity and diversity of identity groups in rural Maharashtra, Shetkari Sanghatana leaders and organizers argue that there is a single issue on which all rural individuals may unite. This is the One Point Plan for remunerative prices. In order to assess its potential to unify, we need to understand rural Maharashtrians' lived experiences relevant to agricultural pricing.

Sociologists Robert Benford and David Snow (1988, 2000) argue that the central diagnosis and objectives of a movement must reasonably match up with participants' actual experience in order to be accepted as valid. This is a condition they refer to as "experiential commensurability." The concept of commensurability has normally been used a tool for understanding the ideational framing strategies of movement leaders, but it also gives us a way to think about the consciousness of participants. Experiential commensurability suggests that participants are not simply gullible recipients of movement leaders' efforts to invent and consolidate an ideology that can motivate action. On the contrary, it enables us to envision participants conducting a sort of conscious "reality check," comparing a movement's propositions with their own actual experiences and other existing interpretive constructs that they have already perceived as relevant.

Of course, the problem of price—of market remuneration for agricultural investments—is experienced differently by different subjects. As we shall see, however, there are definite points of confluence in rural experiences that make price a reasonable rubric through which a wide range of subjects may interpret and engage in a dialogue on the varying challenges they face. As an entry into this dialogue we will consider the comments of two informants who represent two very different economic and social positions.

The first informant is a 34-year-old cotton grower whom we shall call Annasaheb. He owns a large (thirty-acre) holding in Nanded district of Marathwada region, and is a member of a very well-positioned family of "true" Maratha lineage that has, for several generations, held the position of local village *patil*. Although the position today carries significant symbolic capital for Annasaheb and his brothers, in his father's younger years it represented the real power of a village headman.

> **Author**: All of these people you have just spoken about [local participants in the Sanghatana]...I know some of them, and they seem to be in quite different situations than your own. For instance, your family is patil, and you have much more land than they do. What can you have in common?

> **Annasaheb**: It's not so different as you think. See, when my grandfather divided the land, my father and his brother each received 110 acres. My father's share has been shared among my brothers, and now I have 30. My father worked the land all his life on that large field, and he was always in losses. He was always thinking "How come that small shopwalla[1] earns well and I get nothing with so much land?" So, my father tried to make up the difference by lending money or food in the village at a small rate of interest, but he found that he was never able to earn it back because all of the other shetkaris were also in a loss. As a child, I was ashamed that we were not doing better.

> **Author**: And that was your motivation for joining with the Sanghatana?

> **Annasaheb**: No, it was my father's idea. About thirteen years ago, we heard Sharad Joshi speak. He said that the shetkaris were not in a loss because of bad rains, or because they were stupid, or because that is just the way that agriculture is....He said the shetkaris were in a loss because the government is looting them with a negative subsidy that prevented them from getting a fair return. Then my father said to me "OK, we will join these people." And we did.

Annasaheb, whom I had come to know well during my fieldwork, represents a transitional generation from an older order based on heritable titles and privilege to a modern environment of economic competition for status and authority. In many respects, he is the archetypical, cash cropping, capitalist farmer that we might expect to have a central interest in market prices. His story reveals a loss of pride for his family extending from at least two sources: an inability to win much greater rewards from the land than other common agriculturalists, and a sense of antagonism between an old agricultural elite and a rising class of town-centered merchants and traders who are divorced from the soil.

But despite his family's place of privilege in the village hierarchy, Annasaheb's story also points to some sense of identity that he and his father felt with other agriculturalists in the village. His father's inability to reap interest on loans to other (certainly less advantaged) agriculturalists shows a recognition that the lack of profitability in agriculture was a challenge that affected everybody within the agricultural economy. Hence, while Annasaheb generally spoke to me about his interests in terms of the maintenance of his family position and competition with others in the village, he also described his interest in the Sanghatana in terms of at least one point of intersection between his own interests and those of smaller agriculturalists—and this point of intersection was the amount of money remaining in the hands of all agriculturalists after harvest and marketing.

Another example may help to show the relevance of the price issue from a far less advantaged point of view. In 1998, at the tail end of the monsoon season in Western Maharashtra, a stretch of unusually damp weather led to a crisis for cultivators of certain crops in the region, particularly peanut, potato, onion, and tomato, all of which were dependent for growth or storage on what was normally drier weather at that point in the crop cycle. In a village in Western Maharashtra, I spoke with a middle-aged Dhangar-caste couple I'll call Lakshman and Parvati on their two-acre peanut field. As marginal plot holders and members of the Dhangar community—hereditary sheep herders, settled in pockets across the state and usually considered to have a social status at the low end of the Maratha-kunbi caste cluster— Lakshman and Parvati have never known the power and privilege of a village elite. Lifting a tarp of black plastic sheeting and old gunny bags, Lakshman reached underneath and produced a handful of damp, white-looking peanuts. Opening one up, he described their situation:

> **Lakshman:** You see this? There is the beginning
> of a fungus inside. These unseasonal rains will be
> a big problem for us because we cannot keep the
> peanuts dry before we take them for selling. Like
> this, we will never get sufficient money for them.

I asked Lakshman if there was some better way to protect the crop from the rain and humidity, and he explained that the solution was a simple one: a small, covered but well-ventilated storage area. The only problem was that they could never save enough money to build the little structure that was required. At that point Parvati gestured toward the plastic covered mound and said, "This is why we are with Joshi Saheb [Sharad Joshi]." Later in the same conversation, I asked these two if others in their village were participating in the Shetkari Sanghatana. "Some are, and some aren't," Lakshman said, "and there are different reasons for that. But all of us are concerned about prices."

Undoubtedly, there are many other interests and concerns on which these two informants would diverge. In many contexts of opportunity, the patil's family and the Dhangar couple would find themselves at odds in their mutual pursuit of livelihood. However, in discussing their concerns and their motivation for siding with the Sanghatana movement they all describe a degree of shared space that aligns them on the objective of remuneration. Moreover, this shared interpretive space not only tethers each of them to a common ideological space, but also establishes a common communicative space in which these individuals, so unalike in other ways, can discuss personal, local, national, and even international issues with substantial degrees of shared terminology and agreement. But are these informants representative of a larger population that we could call "Sanghatana participants"? Contrary to suggestions that Annasaheb is more representative of the Sanghatana participant base than are Lakshman and Parvati, my field interviews and surveys from around the state suggest that *both* sets of informants are representative of participants in the movement.

Participation in the "one point" movement
Assessing the participation of diverse social segments is one of the greatest challenges to understanding a large-scale movement. Indian sociologist D. N. Dhanagare (1994) has suggested that there are two ways to determine participation. One is to reason backward from the central objective expressed by movement leaders in order to hypothesize about who the beneficiaries of such an objective would be. The other way is to actually "count heads." The challenge here is that, in the case of a large-scale movement, researchers rarely have the opportunity to pursue the latter approach. But in the absence of such an opportunity, can we assume that the other approach—reasoning backward—is sufficiently accurate?

As I have discussed, most observers who have reasoned backwards from the Sanghatana's One Point Plan have concluded that its actual participants comprise a narrow (rich or middle) section of rural society (see, by way of comparison Brass 1994a,

2000; Dhanagare 1994; Gupta 1997; Lenneberg 1988; Lindberg 1995; 1988; Rao 1996). My experience, on the other hand, has been that the participation of both classes of agriculturalists represented by these informants is high. In fact, with the exception of the fully landless—whose participation, though visible, is not nearly proportional to the population of landless labor in the state—participation among poor and marginal agriculturalists such as Lakshman and Parvati is at least proportional to their numbers and, in many settings, it is proportionally greater.

Let us start with some official statistics. According to the Government of Maharashtra Department of Agriculture (1993), land tenure in Maharashtra breaks down into the following percentages:[2]

Marginal Holdings (2.5 acres or fewer)	8%
Small Holdings (2.5 to 5 acres)	19%
Semi-Medium Holdings (5 to 10 acres)	28%
Medium Holdings (10 to 25 acres)	33%
Large Holdings (25 acres or more)	12%[3]

By this breakdown, we see that 20 percent of rural Maharashtrians can be classed as small and marginal holders. At the opposite pole, 12 percent can be classed as large landholders—those who are roughly the class of relatively rich farmers with land and capital that enables them to focus primarily on production for the market rather than for consumption.

How do these proportions compare with my "head counting" of participation in the Sanghatana? In 1996, I conducted a survey at a massive rally for higher cotton prices that the Sanghatana staged in the Vidarbha region.[4] Working with two research assistants in order to cover each of the three entry points used by rally attendees, we randomly surveyed two hundred attendees on a number of indicators, including crops grown, size of landholdings, and the presence and quality of irrigation on their land.[5] The average size of landholdings among our respondents (including both cultivated and uncultivated land) was just over fifteen acres. This is about twice the average holding for the Amaravati administrative division[6] in which the rally was held, and so it does indicate the presence of many large landholders. However, out of this sample 44 percent of the respondents claimed less than ten acres—putting them in the landholding category of semi-medium and below, as determined by the state government census report—and 18 percent claimed five acres or less, distinguishing them as small or marginal holders. In this instance, the proportion of small and marginal holders represented at the rally was just two percentage points less than their overall representation in the state as given in the agricultural census.

As another example, we can consider a much smaller gathering in 1997. In November of that year I attended a road-blocking agitation along a highway in Aurangabad district of Marathwada, and conducted a survey of the sixty participants— people from neighboring villages who were engaged in stopping traffic and handing out to drivers leaflets that described Sanghatana price demands. Of this group, nearly two thirds fell into the category of small or marginal landholders.

These findings point in a very different direction than does reasoning backwards from the One Point Plan. Of course, surveys are not flawless: they can offer some limited insight, but cannot be expected to lead us to a deep understanding of sociocultural experience.[7] For that, we need to go beyond both headcounting and backward reasoning.

Price as a metaphor for outsiders' control on rural livelihood
Let us assume for the moment that the motivating concern of these small or marginal landholders is price. How can we account for this? Part of the answer may lie in the fact that, as I have discussed, Maharashtrian agriculturalists—even those with small or marginal amounts of land, like the peanut growers Lakshman and Parvati quoted above—have an unusually high tendency to participate to some extent in the market (Lenneberg, 1988; Omvedt, 1994a; Varshney, 1995). Given this relatively high representation of people who participate in the market in some

way, it is reasonable to expect that rural Maharashtrian producers would have a higher propensity to agree with each other on the issue of market returns for production. This may also partially account for why "price" is articulated as a focal issue of the Shetkari Sanghatana but not of the major "new farmers' movements" in other Indian states.[8]

But it is also important to consider ways in which the One Point Plan reflects specific, actual experiences of market participation for many producers in Maharashtra. While Sanghatana leaders and participants often express enthusiastic criticism of the Agricultural Pricing Commission (APC) for artificially suppressing prices, the APC is a relatively abstract and distant notion for most producers. Most likely, if it were not for the Sanghatana's propagandizing about the effects of the APC, awareness of this national regulatory body would be limited to only a small handful of highly literate and capital-intensive agriculturalists. However, there are numerous other regulatory structures that a wide range of Maharashtrian agriculturalists experience more directly in their daily economic lives.

For instance, market participation in Maharashtra is subject to state-defined "zonal restrictions." These restrictions are applicable to various crops and prohibit Maharashtrian producers from selling their harvest outside of the state, or outside of a particular designated production zone within the state, regardless of the prices buyers beyond these boundaries are offering. These are especially well known to agriculturalists who inhabit areas near the border zone and see a different price for the same produce on the other side of an artificially drawn line. In the case of cotton, for example—the most important crop in Vidarbha and Marathwada regions—zonal restrictions mean that a harvest can only be sold to the Maharashtrian Monopoly Cotton Procurement Scheme in the grower's specific region, and at the monopoly's own procurement price, rather than to any private buyers or other state's scheme. In the case of sugar, the regions are divided up into factory zones and producers are required to sell the bulk of their cane only to the cooperative sugar factory in their particular zone.[9]

Such regulated sales are not limited to major cash crops such as cotton and sugar. Many of the more pedestrian crops that fall outside of these procurement restrictions—especially storable foodgrains such as rice—are subject to various levies imposed by the state that compel producers to sell a set proportion of their total harvest to the state, at a low fixed price. These levies are not only the principle source of staple grains for the government's redistribution of food through the Public Distribution System (PDS), but also provide the key mechanism by which Indian states control prices and the flow of grains in the overall domestic market by buying and destroying "excess" produce to avoid a market glut. Hence, when the harvest is bad, cultivators are compelled to sell a share at the low, fixed levy price in order to fill the government stores as a bulwark against famine, but when the harvest is *good* and there is a risk of glut, they are compelled to sell an even greater share at the same fire sale prices. This is a dilemma that Sharad Joshi often describes to his audiences as "starve the shetkaris when there is a shortage, and loot them when there is plenty."

These direct experiences of state administered regulations and restrictions, even though some of these are not unique to Maharashtra (such as the levy on foodgrains), have led many producers to understand their situation as a specifically Maharashtrian one, requiring specifically Maharashtrian solutions. The Sanghatana, as a movement conducted in the "language and ideas" of the state, is better equipped to respond to that impression than the conventional movements of the Left that have typically offered pan-Indian solutions.

But such direct experiences of state control on marketing potential do not end there. The state also imposes restrictions on the roles that producers can play in the larger food production process. Due to a host of licensing restrictions (the basis of what is often denounced by Indian economic liberals as "the license-permit *Raj*"[10]), the vast majority of primary producers are prohibited from processing their raw produce into value-added products. This means that, for example, an enterprising cotton growing family cannot legally gin their own cotton to improve their income potential. Likewise, rice growers cannot husk their own rice, fruit growers cannot make their own jams and juices for market, dairy producers cannot process their own milk, and peanut-growers—such as my informants Lakshman and Parvati—do not have the option to convert their harvest into a shelf-stable product such as peanut butter without the requisite licensing. The compounded effect of these policies is an impression among a wide range of Maharashtrian agriculturalists that the state has become an impediment to development and social change rather than the champion of these (see Kothari 1984, 1986; Omvedt 1993, Rudolph and Rudolph 1987). Sanghatana leaders, aware of the crisis of credibility that surrounds these programs, often invoke a paraphrased quote from Mahatma Gandhi, saying "All that is necessary to help the poor is for people to get off of their backs."[11] Joshi's commonly heard position on the matter is recorded in his Sanghatana training manual: "Nothing needs to be done to remove poverty. Simply put an end to the tremendous effort the government exerts to sustain and increase poverty, and it will be enough to stop right there. The poverty will eliminate itself" (Joshi 1988, 76).

While Sanghatana participants broadly express frustration with any of these programs they have personally experienced impinging on their own economic choices, Joshi and other Sanghatana leaders have helped to build social unity out of that frustration. They do this in part by educating participants on each of the other programs that affect other participants, other types of agricultural producers, and other regions of the state in similar ways. They also provide a logical critique of these programs through a standardized lens of operational costs and expenses. Joshi has developed a rich new vocabulary for some aspects of his critiques. One of many neologisms in the Sanghatana lexicon is *negative subsidy*. This refers to the gap between the positive benefits derived from state subsidized inputs and the losses that agriculturalists incur due to price suppression and other market-regulation policies. Negative subsidy stands for the antithesis of fair or market-based pricing, and it signifies the exploitative status quo that, according to Joshi, affects all agriculturalists.

This particular perspective—that a core set of policies negatively impacts rich and poor agriculturalists alike—is significantly distinctive both regionally and historically. Although there is some evidence that pricing may have been a focal point of collective agrarian action in Maharashtra's past (see, for example, Charlesworth's analysis of the Deccan Riots of 1875[12]), the limited conventional agrarian organizing that has taken place in twentieth-century Maharashtra has been overwhelmingly oriented toward land redistribution, debt relief, and government subsidization of inputs, rather than remunerative prices (Rodrigues, 1998). Expressions like negative subsidy[13], *vajib dam* (fair price), and *ghamace dam* (return for our sweat) serve two simultaneous purposes: they provide vocabulary for a distinctively "Sanghatana" language of discussion on the causes and solutions applicable to rural poverty, and they constitute an ideological negation of any other movements and politicians that advocate expanding the state's subsidies or other interventionist programs. The movement adds further force and credibility to this language by supporting its claims and critiques with rational analyses that tack back and forth between the experiences of individual agriculturalists and regional, state, national, and even global comparative statistics.

For example, when explaining the meaning and effect of negative subsidies at activist camps and mass rallies, Joshi (who has a background in accounting) explains that he has carefully calculated the overall value of existing government subsidies on key agricultural inputs in relation to the agriculturalist's actual lost revenue due to price controls, zonal controls, levies, and other restrictions. In this way, he has determined that the net balance represents an overall loss to the producer of negative 2.33 percent—in other words, the *negative* in the subsidy. This gives a rational and scientific character to the movement's claim that subsidies and handouts are little more than bandages or false fronts for a system that is inherently exploitative. Negative subsidy is a term widely known and used among Sanghatana activists and general supporters who, citing Joshi's calculations, declare Maharashtra's to be the *highest* negative agricultural subsidy in the world. By contrast Joshi has calculated that farmers in the United States enjoy a *positive* subsidy of 26.17 percent while farmers in the European Economic Community (EEC) enjoy a positive subsidy of 37 percent.[14] A surprising number of participating villagers around the state can cite these figures, repeating what they have heard from Joshi and other Sanghatana leaders.

The question of labor and non-agriculturalists
The foregoing gives credibility to the idea that rich and poor agricultural producers could find some common ground on the issue of government pricing and production control. What about other rural Maharashtrians who are not market-oriented producers, such as agricultural laborers or people in non-agricultural rural occupations? Although it is clear from my research that these categories of people are not heavily represented in the Sanghatana, there is definitely a large enough presence for us to wonder how, if at all, the Sanghatana's central demand could be of relevance to their interests. Several factors can help us make sense of this.

A very important factor, with regard to labor, is that much of the Maharashtrian agricultural labor force is actually *cultivating labor* rather than fully *landless labor*—in other words, households that practice some cultivation for the market but also sell labor as a supplement to their income. These types of households, which identify with labor in some contexts and with landed agriculturalists in others, are particularly difficult to count.[15] From my fieldwork, I have observed that there are a large number of households that generate their livelihoods in this class-indeterminate (and often *caste*-indeterminate) space, and that many households shift back and forth between being cultivators and being laborers from season to season, year to year, and through different stages of household evolution (see also Carter, 1998).

Census data also give us some insight into households in the gray area between cultivating and laboring. Although the agricultural censuses suggest a progressive rise in agricultural labor as a percentage of the rural population in the state—accounting for 27 percent of the total workforce at the time the Sanghatana emerged in 1981 compared to just 23 percent two decades earlier (Brahme and Nene, 1985)—it is noteworthy that the most significant rise seems to be not in laboring households that own no land but rather in households that own too little land and must work to supplement their resources (see Omvedt 1994b). In other words, the ongoing process of land fragmentation from generation to generation—along with other rural economic challenges—has pushed increasing numbers of families into the small and marginal category where, in addition to selling their labor in order to make ends meet, they may continue to produce crops for the market. Some of these households may be more dependent on wages (hence consumers interested in lower prices) and others more dependent on market production (hence producers desirous of higher prices) depending on their situation at any point in time—but if and when any of these households have a sizable share of marketable produce alongside their production for household use, profitable pricing could be perceived as a logical objective.

For these landowning laborers—as well as for landless laborers—wages are also an issue of tremendous concern. Here it is important to consider efforts the Sanghatana has made to link this wage concern with the overarching objective of higher prices. Joshi, who has frequently decried the government's established agricultural minimum wage figure as "absurd" and "fit only for animals," regularly implores the labor-employing participants in the movement to voluntarily pay a higher day-rate to their laborers. And, though it may seem to the outside observer merely a distant hope that the benefits of higher prices will trickle down to labor, it is meaningful for laborers in the movement that many—though certainly not all—landowning Sanghatana participants (including Joshi) who employ labor have made good on this voluntary increase. It is also meaningful that, under Joshi's influence, nearly all the other price-oriented movements within the Interstate Coordinating Committee (ICC) have begun to include higher minimum wages for agricultural labor in their own calculations of local production costs. In other words, the Sanghatana's demand for remunerative price *includes* a demand for living wages for laborers—and this is recognized and widely discussed by laborers and landowners alike.

The result of these efforts is that the Sanghatana has established a perceivable track record of sincerity in connecting wages to the remunerative price issue, and has facilitated at least some points of intersection between the interests of agricultural producers and agricultural laborers. I have met some rural labor organizers outside of the Shetkari Sanghatana who have perceived the Sanghatana leadership to be sincere in their call for higher wages. One such person, who I shall call Rajiv, is a civil rights leader involved with a wage-oriented movement of landless adivasi and Dalit laborers. Telling me about an encounter that his organization faced with a local reactionary landowners' group, Rajiv described how he first learned to appreciate the position of the Shetkari Sanghatana:

> **Rajiv:** A few years ago, some people in our district—they were all big landlords, former keepers of bonded-laborers, and very caste-ist and conservative—started their own organization to resist our labor unionization movement. They called themselves the "Shetkari Sanghatana." We were concerned that they were attempting to gain legitimacy by using the very same name as [Sharad] Joshi's movement. So, we contacted Joshi about this, and he said he would make a point of publicly endorsing the work that we are doing to unionize the laborers. He also said that he would call upon the landlords' group to accept our proposals for better pay and conditions....I still don't completely agree that

remunerative pricing is a solution to the laborers' problems, but we have decided that Joshi is sincere about this.

Another important factor in understanding any potential appeal of the remunerative pricing issue for agricultural laborers, is that market participation is aspirational. Laborers with no land aspire to acquire some land and sell produce. Laborers who already have a bit of land and sell some produce aspire to sell more. In another conversation, Rajiv described cultivation as an aspirational identity among the laborers in his movement:

> **Rajiv**: Look, it's like this...every villager wants to be a cultivator rather than a laborer. Even someone who rents or owns just a tiny piece of garden plot will still prefer to identify himself as a landed farmer—and will exercise that identity in social and political realms.[16]

For families that have decreasing access to land due to fragmentation, this aspirational identity represents a strategic social status and historical identity as non-laboring households. For others who never held viable amounts of land, the aspiration is a well-established strategy for elevating social status and for building relationships of relative equality with other members of the community. Moreover, as an aspiration, the acquisition of land is not out of touch with possibility; it is informed by historical realities of land redistribution schemes, successful "takeovers" of public land or landlord-owned land by the landless, and even purchases of land by landless families—all of which have been observable phenomena over the course of the twentieth century (see Attwood 1979; Carter 1988).[17]

Finally, there are some village participants who neither own land nor sell their labor working in the fields of others. That is, people in nonagricultural or semi-agricultural positions. In my field research I have occasionally encountered a variety of such individuals at Sanghatana activities, including barbers, well diggers, cattle herders, and shopkeepers. For most of them, the primary appeal of participation resides in issues other than price, but many have also spoken of price as a significant point of interest. As one participant told me at the road-blocking agitation in Aurangabad district, after I asked him how much land he was cultivating: "I am a tailor; I have no land. But in the village I cannot have a business if nobody has any money."

Individual reasons for the participation of cultivators, laborers, and people in non-agricultural occupations are certainly complex and resistant to generalization. The social rifts shaped by economic stratification and rural hierarchies of power are wider for some than others. This is particularly true for the state's historically disadvantaged communities of Dalits and adivasis. Members of these groups—who are more likely to suffer from lack of land, impingement on their customary uses of public land, and entrenched discrimination—have a disproportionately high representation in the agricultural labor workforce. Even where they do have some land, it is most often land of the poorest soil and terrain while Marathas (even those that have seen their landholdings shrink from generation to generation) tend to own the better quality fields with the greatest potential for per-acre production (Vora and Palshikar 1990). Dominant caste Marathas continue to have disproportionately high ownership of Maharashtra's coveted *bagait* (irrigated) land, which is vastly more productive than the state's majority of rain fed *jirayat* (dry) land.

There are some general conclusions we can draw about the economic appeal of the One Point Plan. Recall that in the previous chapter we saw some of the wide economic variations across the state and noted two areas of the state where the Sanghatana has had negligible organizing success. One is the coastal Konkan region. In the Konkan, where cash crops are a minor component of the rural economy, state-run marketing mechanisms and state controls on pricing are a significantly lower burden for agrarian households. Moreover, rural households in the Konkan tend to be more integrated in the urban economy of Mumbai, often drawing a large percentage of their household income from the urban labor remittances of family members. The other mobilizational void is in the far east, in the heavily adivasi populated districts, where rural economic issues are largely focused on forest products and adivasi rights of access to forestland.

Conversely, the areas where the Sanghatana *has* had consistent mobilization success are those that have a high level of cash cropping and have seen the lowest levels of state-sponsored rural development and of industrial wage opportunities. These are the characteristics of the majority of the districts in Marathwada and Vidarbha, which are precisely the regions where the Sanghatana has been most successful.

Communicating experiential intersections through Maharashtrian signs

All of this goes a long way toward understanding the observable unity of the Shetkari Sanghatana and its capacity to attract a broad socioeconomic cross section of participants—but it can only provide part of the answer. Even in geographic areas where the Sanghatana's focus on price may make more sense, the challenge remains to understand how the Sanghatana has built a community that crosses other deeply felt identity groups and other lines of rural social and economic differentiation—particularly those correlated with caste. As we discussed in the previous chapter, this challenge is in many ways greater in Marathwada and Vidarbha, where caste and religious conservatism are stronger than in other parts of the state. Here we must consider the Sanghatana's simultaneous embrace of these differences *and* its ambiguation of these differences through other, noneconomic elements of its interpretive idiom.

Political scientist Ashutosh Varshney (1995), writing about the difficulty of uniting subjects under a strictly economic model of mutual interest in Indian rural movements, has argued that, "the ultimate constraint on rural power...[may] well stem from how human beings perceive themselves—as people with multiple selves." Varshney concludes that, though the potential for rural economic unity in India is great, a "preponderance of the economic over the non-economic is not how this multiplicity [of selves] is necessarily resolved" (4-5).[18] How does the Sanghatana manage to craft a movement identity out of such a wide range of selves? As an entry point into this discussion, we can look more closely at the core ambiguating term in the Sanghatana lexicon, the word that is referential of the community and its participants—shetkari.

"Shetkari" as a sign of self and community

If the One Point Plan suggests, or even reflects, common experience across wide-ranging rural inhabitants, this unity is reinforced through the core term of identity used within the movement. This term, shetkari, is a common Marathi word. It is deeply embedded in the language, but as an identifier of community it has a number of distinctive characteristics. In contrast with common, relatively specific Marathi agricultural occupational terms, shetkari has a much more expansive set of potential references. The Marathi term *kashtakari,* for example, refers to a toiler or agricultural laborer. The Marathi term *bagaitdar* means a cultivator with an irrigated field. These are quite specific. The word shetkari, in its broadest sense, means simply "a person connected to agriculture," a meaning sufficiently ambiguous to embrace nearly all agricultural classes and occupations. Moreover, unlike some other narrower multigroup terms like kunbi or Dalit (the latter meaning "downtrodden") shetkari is agnostic of acquired caste and caste-based social, cultural, and political implications.

Shetkari, then, is what we could consider (within the ambit of rural society) an open signifier. This makes it an ideal term for expressing the movement's inclusivity of rural social groups— and it is able to do this at a level of abstraction that does not threaten other constructions of identity. At the same time, it expresses differentiation, which is crucial for any sense of identity. One way it achieves this is through opposition to non-rural communities and non-rural geographies. In the idiom of the Sanghatana, these oppositional Others are the "black British" and "India."[19]

Within rural society itself, shetkari also contrasts with another term that, prior to the Sanghatana's rise, had most frequently been employed in agrarian organizing in Maharashtra and throughout most of India. This word, *kisan,* is a Hindi word for "agriculturalist" that originally may have had much of the same ambiguous scope. However, recurring usage by left-oriented movements throughout the state and country over the last century (and continuing today), has given kisan a signification that is decidedly class-oriented toward smaller cultivators and the laboring rural poor. Through this usage, it has also acquired association with conventional redistributive (rather than price-oriented) peasant movement agendas, including "land to the tiller" struggles and subsidies for the agrarian poor—which are precisely the conventional objectives opposed by the Sanghatana. Thus, shetkari embodies much greater potential for an ideology that postulates the unity of all agrarian interests and the primacy of the price objective.

Another reason that shetkari is such a useful term is that, as a Marathi (rather than Hindi) word, it has the capacity to convey an identity unique to the Maharashtrian agrarian experience. This differentiates the movement from pan-Indian movements, and also establishes a degree of continuity with other specifically Maharashtrian struggles. It is important to remember that the rural economy and agricultural ecology in Maharashtra—particularly in the highly mobilized eastern regions of Marathwada and Vidarbha—is materially distinct from that of most other Indian states. As mentioned above, marginal land-holding Maharashtrians have a greater propensity to participate in the market than agriculturalists in other parts of the country. At the same time, poor soil, variable rains, and uneven development of irrigation in Maharashtra make agricultural productivity less dependable than in neighboring states. This distinct ecological situation is well understood by Maharashtrian agriculturalists, and lends popular credence to an idiom that seems to reflect specifically Maharashtrian issues rather than one with pan-Indian implications.

Of course, simply employing a term that has the capacity to embrace multiple groups is not sufficient to make the unity of these groups a credible proposition. As we have discussed, there are definite confluences of rural economic experience that may impact the lives of diverse subjects in related ways, but these do not necessarily lead to a shared understanding of experience or a shared desire to mobilize in a common interest. How does the Sanghatana render the idea of unity credible? In other words, how does it manage to correlate its interpretive framework with such a wide range of on-the-ground experiences and perceptions of opportunity? One way the movement approaches this is to define shetkari unity as situational, as an abstraction that *embraces* rather than *erases* the range of subject positions and identities among its potential participants.

"Shetkari" and multiple selves

It is clear that a simple issue like price—even when this encompasses related issues like wages—cannot account for all constructions of interest in rural Maharashtra. Participants in the Sanghatana rarely deny that there are other valid interpretations of the problems they face and other opportunity structures that benefit them in other contexts. Furthermore, many participants who struggle side by side in the context of the Sanghatana often find themselves opposed to fellow *sanghataks* in other contexts of social action and expressed identity. Therefore, because the Sanghatana cannot represent all rural interests in all contexts, a great deal of its credibility extends from recognizing the situational nature of shetkari unity.

This situational unity is clearly expressed by Sanghatana leaders and activists. We can see an example of this in a lecture by Sharad Joshi. Speaking at a two-day training camp for Sanghatana organizers, conducted in the village of Ambajogai in 1981, Sharad Joshi described the logic of shetkari unity in the following conditional terms:

> In order to understand the concept of Bharat and India we must think of all the people in the rural

areas as being similar. In reality, we know that the people who live by agriculture in the rural areas represent many different types. There are cultivators having large, irrigated lands, cultivators with small but irrigated lands, cultivators with large holdings of dry land, cultivators with just small patches of dry land, agricultural laborers who also have some very small land of their own, and laborers who are completely landless.[20]

In the lecture, Joshi then went on to explain that, in the context of the struggle against "India," the differences are less significant:

The argument is commonly made, especially by the leftists, that the shetkaris cannot possibly be considered as one. We don't say that there is no difference among all these types of shetkaris. Obviously, there is a big difference between large land-holding shetkaris and smallholding shetkaris and the laborers who have little or no land at all. Likewise, there is a big difference between the Tatas or the Birlas[21] and their junior officers and the workers in their factories. In both societies—Indian and Bharatiya[22]—there are different levels, but the line of conflict that we have drawn, the line between India and Bharat, is at the point where the differences are the most important. This is the line across which the greatest exploitation takes place...poverty in Bharat results from the fact that nobody there gets sufficient compensation for his or her work.

This proposition of a conditional unity among diverse social groups is often considered to be something novel in the Shetkari Sanghatana. In fact, it is not that unusual. If we recall other historical movements in Maharashtra discussed in chapter two—the Varkari Sampradaya pilgrimage cult in the thirteenth century, the ideological and political struggles for Maratha encompassment of multiple caste groups, the united anticolonial struggles, and the proposition of pan-Marathi collective selfhood in the Samyukta Maharashtra movement, to name just a few— then we know of many past performances of conditional mass unity in pursuit of a cause.

Each of these historical movements entailed, at least in certain aspects of their expression, propositions of mass unity juxtaposed to a dominating Other—whether a regime of "outsiders," or a dominating community of insiders. The Shetkari Sanghatana's leaders have made extensive use of these past Maharashtrian movements as symbols to reflect the Sanghatana's own conditional unity. For example, the Sanghatana's first major effort to take the message of the movement on the road—essentially a propaganda tour through rural areas of the state—was represented as a shetkari *yatra* (pilgrimage) to Pandharpur, following the caravan of the Varkari pilgrims on their annual march.[23] The intention in this is clear: to align the Sanghatana with the Varkari's own rural ethos and the Varkari Sampradaya's assertion of rural unity against mutual oppression. Similarly, the Sanghatana's efforts to signify king Shivaji as a rural benefactor and a paragon of tolerance can be seen as a move intended not only to discredit the interpretive complexes of Maharashtrian nationalism and Hindu chauvinism, but also to align the Sanghatana with a specific core popular meaning of Shivaji: unity in the face of outside authority.

Thus, when invoked in the appropriate contexts, shetkari is an excellent umbrella term for a broad cross section of rural Maharashtra. However, given that other identities have significance and utility in other contexts, the Sanghatana also needs to embrace rather than attempt to override many of these other identities and interpretations of experience. This is crucial to understanding why the (otherwise) scientific and economistic language of the One Point Plan has been co-signified through many of the same historically deep symbols with which other Maharashtrian identities have been constructed and maintained.

Organizing in the "language and culture of the state" is not only a strategy aimed at achieving the broadest possible resonance, it is also a strategy for going beyond this, claiming signification rights to resonant signs and inserting new meanings on those signs that highlight the economic points of intersection along which the Sanghatana organizes.

Shetkari and price in the field of signs
The strongest idioms of movement unity invoked by Sanghatana leaders and participants are those tied to remunerative prices and the restoration of Bali's mythic kingdom, but these are by no means the full collection of popular signs invoked in the movement. The range of signs that the Sanghatana has employed to craft the movement's meaning can be readily seen during a visit to Angarmala, the Sanghatana's headquarters described at the beginning of this chapter.

The walls of Angaramala bear recruiting and propaganda posters that span twenty years of Sanghatana events and political agitations. Many of the posters outline specific demands associated with each of the major campaigns and rallies for which they were created. In addition to the recurring theme of fair prices (*vajib dam*) and "return for our sweat," they also call for Bharat's independence (*swatantrya,* with its ever-present implication that this may apply not only to rural areas in general but also to secessionist regions in the state); economic and social liberation for women (*stri mukti*); liberation from debt (*karj mukti*); fair agricultural labor conditions and wages (*ucit majduri),* and they link different campaigns to price policies on cotton, wheat, milk, or jowar.[24]

But the posters are also an amalgam of ideas that include not only the Sanghatana's economic diagnosis and specific demands, but also symbols and ideas of other Maharashtrian identity groups and an assortment of deeply Maharashtrian cultural signifiers. Many of the posters contain images of historical and mythological figures that suggest the movement's efforts to ally itself with past ideals, embrace a range of rural identities, and reclaim these iconic figures from their use by other divisive or oppositional groups. In one poster, there is an image of the Varkari deity Vithoba. The accompanying text asks pilgrims heading to Pandharpur to demand that the god grant the shetkaris freedom from oppression. In another poster, the "shetkari king" (*shetkaryanca raja*) Shivaji is depicted side-by-side with the twentieth-century Mahar Dalit organizer and Buddhist convert Dr. B. R. Ambedkar, in a clear attempt to wrest Shivaji away from the discourses of Maratha-caste greatness and Hindu nationalism.

Elsewhere, the anti-black British struggle of the Sanghatana assumes continuity with the national freedom struggle against the (real) British. In this poster, a man named Babu Genu, one of the first Maharashtrian freedom fighters killed in the national anticolonial struggles, is depicted being crushed under the wheels of a colonial British truck. And, of course, there are also images of the good demon king Bali, and of Sharad Joshi addressing crowds. Throughout the room there are nameless heroes as well—pictures and drawings of proud shetkari men and shetkari women wielding plows and sickles. Notably, some of the men wear turbans, some wear the *topi* cap of the Congress Party, and some sport the unmistakable beards of a Muslim or Sikh. All of them are shetkaris.

The campaign posters depict images of burden, conflict, and struggle, clearly linking these with the Sanghatana cause. In one, a woman is hauling a headload while carrying an infant on her back; the Marathi slogan reads "how much can we endure?"[25] In another, a menacing tiger (the symbol of the Maharashtrianist and Hinduist Shiv Sena party) bears down on a shetkari from one direction, while a clawed black hand grasps at him from the other. The shetkari, whose mustache and garb are reminiscent of images of Bali, holds up the Sanghatana logo as a shield in defense, beside text that warns: "while escaping the clutches of the black British, don't get caught in the jaws of caste-ism/communalism."[26] And in another poster, the humble Bali himself is depicted being crushed into the earth by Vamana the dwarf, beside the Marathi words "Be alert! Be alert!! Will the caste-ist/communalist

'Vamana' once again quash the kingdom of Bali?"[27] In some posters, the kingdom of Bali is signified as something that has been partially reestablished and in need of protection; in others, it is signified as something that is "on its way." In the former case, Bali's kingdom is decisively associated with the Sanghatana itself and the mobilization of the rural masses. In the latter, it is unmistakably a metaphor for the eventual realization of the Sanghatana's goals.

Photo 3. A Sanghatana poster with an image of Vithoba. The text asks pilgrims to challenge the deity to improve the condition of the shetkaris.

Glancing through the room on my first visit, I was struck by the range of the "language and ideas of the state" the Sanghatana has called into service, and the extent to which the strict economistic language of the Sanghatana is interwoven with symbols that had been shaping and reflecting discourses and identities for centuries before the Sanghatana ever emerged. This same mélange of symbols, slogans, and ideas can be found in almost every setting where the Sanghatana is performed and crafted. It is in the writings and speeches of Sharad Joshi and other movement leaders; it is on the banners hung and in the songs sung at Sanghatana rallies; and it can be heard in the talk of district activists and village participants.

Photo 4. A Sanghatana poster with claws representing the black British and a tiger representing communalism. The text urges shetkaris to vote for non-communalist parliamentary candidates.

Combined with the economistic agenda for higher prices, these Maharashtrian signs constitute a culturally embedded set of ideological associations and propositions. Collectively, they establish the frame, or interpretive scheme, through which the Sanghatana identifies social problems, diagnoses their causes, prescribes solutions, and motivates action. This framing is more than a simple set of economic (or social, or political) diagnoses and demands. It is also a sort of cultural wrapping that selectively links a demand, or set of demands, to a broader set of existing values, cultural ideals, norms, and attitudes.

Earlier I pointed out limitations in the way many scholars have theorized the creation of social action frames—specifically, their emphasis on the agency of movement leaders, often overlooking the contributing productions of movement participants. But the literature on social action frames nonetheless helps us understand some of the ways communities attempt to build shared interpretive schemes that add emotional richness to otherwise strictly rationalist intersections of material interest.[28] An effective interpretive scheme can help to create a *culture* of shared experience that has some potential to bridge rural subjectivity across what Varshney (1995) calls the "multiplicity of selves."

One of many valuable notions from frame theory is what Robert Benford and David Snow (2000) call *narrative fidelity*—or the extent to which proffered framings "resonate with the targets' cultural narrations" (622). Narrative fidelity fails when cultural symbols are drawn into a framing of the world that unacceptably violates interpretations that people have already internalized and embraced. This means that an ideational frame must not only credibly correlate with participants' experiences of livelihood and of the movement itself, but must also be reasonably consistent with existing cultural meanings. This is complicated when the popular symbols and narratives of the Marathi-speaking regions have been deployed for so many diverse purposes by so many diverse selves. If Shivaji, for example, has popular circulation as a symbol of Maharashtrian nationalism, or Hindu or Maratha supremacy, how can the Sanghatana credibly invoke this same symbol as a signifier of ruralism and rural caste-communal unity?

Photo 5. *A poster spreads news about a Sanghatana rally, jointly commemorating the birth anniversaries of the seventeenth- century king Shivaji (left) and the twentieth-century Dalit organizer Dr. B.R. Ambedkar (right). The Shetkari Sanghatana logo is in the center.*

Clearly, the use of the "language and ideas of the state" entails much more than merely co-opting existing ideas and beliefs. Signification of the movement through these multivocal vehicles must also attempt to transform the symbols themselves through new sign-associations and selective spins on meaning, maximizing their capacity to signify the movement and correlate with experience to the greatest effect. In my understanding, the Sanghatana's symbolic complex works to maximize the mobilizing effectiveness of these cultural resources in several ways. Specifically, through significations of inclusivity, continuity, and differentiation.

Signifying inclusivity, continuity, and differentiation
One way that the Sanghatana's complex of signs maximizes the effectiveness of the resources employed within it is by combining

signs in novel ways to create meanings of *inclusivity*. By drawing on a wide range of symbols that are deeply associated with different Maharashtrian identity groups, the Sanghatana's symbolic complex implicitly defines the range of rural subjects that the Sanghatana claims could benefit from the demand for higher prices—essentially anyone who lives in the rural areas, whether Maratha or Dalit, Hindu or non-Hindu, male or female. A second way is employing signs and spinning meanings to express *continuity*. This is done by drawing on symbols of past struggles in the state and linking those struggles in selective ways to the Sanghatana. The effect is a signification of historical continuity with other social issues (for example, the women's struggle, Dalit rights, religious equality), other movements of liberation (the liberation of the Marathi lands under Shivaji, or Maharashtrian efforts in the national freedom struggle), and other expressions of mass resistance to domination (the Varkari pilgrimage cult, or the Nonbrahmin veneration of Bali in the Satyashodhak movement). The third way that effectiveness is maximized is by linking signs with meanings that express *differentiation*. By selecting specific symbolic resources and emphasizing or negating key aspects of their popular meanings, the Sanghatana differentiates itself within a larger field of competing ideologies and interest groups—particularly those with mobilizing ideas and objectives that are contrary to its own goals, such as conventional redistributive agrarian politics and the ideologies of caste-ism, Hindu chauvinism, and Maharashtrian nationalism.

Each of these objectives—inclusivity, continuity, and differentiation—are of critical importance for a mass movement in a social context characterized by diverse histories, diverse identity groups, and competing claims to represent different target communities. These objectives have an impact not only on the movement's invocation of historical figures, mythological narratives, and other movements or social issues, but also on the routine, vernacular language employed within the movement.

However, while drawing on these cultural resources, the Sanghatana is not just throwing together a little bit of everything in an attempt to accommodate other identities, or establish continuity and differentiation. As each of these cultural signs is drawn into the Sanghatana's ideological complex, they become part of a new constellation of signs in which each component piece becomes associated with other component pieces. Placing Shivaji and Ambedkar together, for example, has the effect of endowing each of these with new propositions of meaning. Constellated in this way, the Dalit cause, for example, is rendered as a *Maharashtrian* cause rather than the narrow interest of a specific group. Similarly, the ideal of social and political liberation that Shivaji represents is rendered to include the liberation of the downtrodden and non-Hindus (in this case, Buddhist Dalits).

Moreover, these signs also become constellated with the meaning of the movement itself, along with its diagnoses and prescriptions. In other words, both Shivaji and Ambedkar become associable with specifically rural issues and the ideal of rural unity and liberation. Likewise, Dalit exploitation (signified by Ambedkar) becomes expressive of rural exploitation in general, just as real liberation of the Marathi masses (signified by Shivaji or the return of Bali) is represented as the liberation of the people of Bharat. In the same way, linking the Varkari Sampradaya into this constellation of signs highlights the elements of rational social criticism embedded in the bhakti movement, and links the pilgrims' narrative of rural liberation to the overtly economic diagnoses and prescriptions of the Shetkari Sanghatana. When all of the signs and meanings included in the Sanghatana's ideological constellation are taken together, the whole communicates the movement's intent to include the entire cross section of rural identity groups, its aspiration to be associated with other progressive causes in rural Maharashtra, and its distinction from rival interpretive schemes that fragment or oppress the rural community.

Core themes: unity and liberation

These ongoing associations and juxtapositions are able, over time, to transform signs and highlight their most appropriate meanings for the movement. But how does such a novel constellation of symbols manage to hang together as a credible whole? To understand this, we have to return to the two key themes that underlie the Sanghatana's entire complex of economic and cultural ideas: shetkari unity and shetkari liberation. These two themes serve to establish the meaning that is common to all other elements of the complex, and they are the benchmark for all other significations of inclusivity, continuity, and differentiation.

To give an example of how inclusivity, continuity, and differentiation align with the core themes of unity and liberation, we can consider Joshi's representation of one important cultural sign, the Indian Independence movement. In the following example, Joshi is addressing attendees at a week-long activist training camp that I attended at Angarmala in September 1998. This excerpt, translated from Marathi, is from one of many classroom-style training modules that made up the camp's program. The subject of Joshi's lecture was the history of Indian social organizing and its relationship to the emergence of the Shetkari Sanghatana.

> **Joshi:** Even during our freedom struggle, we were saying that the whole of Hindustan[29] should get independence. We were not talking only of political independence—we wanted both "self-rule" [swarajya] and "just-rule" [surajya].[30] Even at that time there were differences between the people. There were rich and there were poor. There were differences of caste and religion. Yet, in spite of these differences, the common struggle for independence was waged on the assumption that the main line of exploitation was between Britain and Hindustan. Today we know that the exploitation continues, but the line of exploitation is between Bharat and India. So, just as people then kept their internal differences

aside and participated in the freedom struggle, so
must now be the case for the freedom of Bharat.

Here, as in many other public contexts in which I have heard
Joshi speak and interact with movement participants, the deeply
historic cultural value of *swarajya* (self rule) is invoked with
specific reference to the current struggle of Bharat and the
shetkaris. Although Joshi was speaking specifically about the
freedom struggle against the British, swarajya is also inextricably
resonant of other past struggles, such as the Maratha struggles
under Shivaji and the Marathi statehood movement—and it is
essential to the story of Bali Raja and the overthrow of his just and
legitimate rule. All of these are invoked at once through their
common association with struggles for freedom, and are
resignified in continuity with a specifically shetkari liberation and
shetkari unity in the face of common oppression. The proposition
of continuity suggests that both the national freedom struggle and
the Sanghatana are characterized by a socially inclusive base of
support, but Joshi also differentiates the Sanghatana from the
national struggle by identifying a different "line of exploitation"
that must be opposed.

The effect of such movement narrations is that shetkari unity
and liberation are held up as not only the ultimate meaning of the
Sanghatana and its policy agenda, but also the core meaning of
each cultural sign through which the Sanghatana expresses itself.
And, herein lies the key to the movement's differentiation:
whatever is not conducive to these twin objectives is not
expressive of the Sanghatana. For example, Shivaji is a signifier of
the movement only when the meaning of Shivaji is allied with the
struggle of agriculturalists and their attainment of justice in a free
economy. The Varkari Sampradaya signifies the Sanghatana
movement only when it is understood as part of an egalitarian
struggle to validate the masses residing in the social and
geographic periphery. The national independence movement and
the Maharashtra statehood movement are signifiers of the
Sanghatana's continuity with the past, but only when these are
expressed as unified struggles of the "true oppressed" to shake

off the yoke of rule by outsiders. In the same way, the demon king Bali of ancient myth signifies the Sanghatana only when he is understood as the ideal of swarajya and a victim of deceit by forces that sought to mine his wealth. "Price" fits into precisely this same formula: it is held up as a tool for the liberation for all shetkaris, and an objective that bridges their otherwise disparate social and economic positions. All of these meanings interdigitate with each other, and with the overarching constellation of meanings that is the Shetkari Sanghatana. Any aspects of these that serve the cause of India (not Bharat), the neocolonial black British (not the shetkaris), caste supremacy, religious chauvinism, or rural class struggle (as opposed to unity) are dissociated from the Sanghatana and represented as illegitimate meanings.

Expanding the context of shared experience

Cultural symbols that have broad acceptability and a plurality of embedded meanings have a distinctive utility in mass movement organizing. We would be mistaken to view a movement's invocation of these either as a simple co-optation of "the popular" by the movement's leaders or as the pure and unmediated expression of popular identity by the masses. As I have discussed, movement leaders and organizers cannot construct meaning out of thin air, or simply "market" movement ideas and objectives to a mass audience (see Melucci 1996; Oliver and Johnston 2000). Participant's own felt interests and subjectivities play an important role in the framing and "meaning work" that makes these symbols effective.

Despite this, it is clear that leaders and empowered social groups within a movement have disproportionate access to certain resources in the construction of meaning. These people have much better and more immediate access to apparatuses of signification, greater access to influential social networks, and substantial control over the creation of performative contexts in which participants *experience* the movement ideology and the movement itself. These advantages can be classified in terms of three interrelated capacities: communication, organizing, and action. In subsequent chapters I will consider these dimensions of the movement from a bottom-up and participatory perspective, but at this point it is important to recognize the unique strategies and resources that leaders are able to employ within the broader dialogue on Sanghatana meaning and identity.

Constructing meaning through communication resources

The visible, vocal, overt level of central leadership is the level at which most outsiders perceive the movement. It is also an important—though not the only—level at which participants perceive it. This means that not only is the central leadership distinctively empowered to communicate the movement to its actual and potential following, but also to represent the movement's goals and ideology to political rivals, media, scholars, and other non-movement audiences observing from the outside. Much of the singularity of voice that we tend to perceive in movements extends from this unparalleled capacity of movement leaders to communicate on a broad scale. Not only do leaders have significant ability to define and declare the movement's objectives, tactics, and ideology at any given time, but also to declare the meaning and significance of these in a voice that is more consistent, more prevalent, and more unified than other voices within the movement.

In the case of the Shetkari Sanghatana, there are several different types of resources that constitute this capability. One is its network of village workers who put the word out in a relatively consistent fashion across the state. Another is its corps of central and regional leaders, who travel to rallies and meetings in order to represent the authoritative voice of the movement. And, of course, there is Sharad Joshi, who travels extensively throughout the year visiting villages, conducting training camps, addressing rallies, and supporting sympathetic political candidates. Many activists describe the charisma and high mobility of Joshi and other leaders as fundamental to what one informant—who I will call Pramodh, a well-educated and town-based man who was once a statewide organizer for the movement—referred to as the Sanghatana's "communication miracle."

Pramodh: From the beginning, it was a communication miracle that took place just through word of mouth, spread locally. See, here was a leader [Joshi] who was a brahmin...no farming background, an urban bureaucrat, educated, worked for an international organization, lived abroad, totally westernized lifestyle[31]...all together as unfit a candidate as one could have to become the leader of farmers in Maharashtra. In the early days of the Sanghatana he was ridiculed by observers...they said he wouldn't know how to talk to farmers.[32] But he did some farming, he kept accounts, and he picked up enough of the rural language. No, not the rural language, let us say, because his language is still very urban...but, enough of rural wisdom, let us say. And when he started telling farmers that their poverty was not from laziness, or due to their conventional methods (as outsiders said) or their fate (as farmers themselves felt) but because of a state conspiracy to industrialize and urbanize at the cost of agriculture...when he told them that state policy was "death to the farmers"...when he told them about India vs. Bharat, and drew parallels to Bali and his fate...they understood directly and didn't need to have it explained a second time. Thus was the miracle.

Author: And how did he spread that message?

Pramodh: By going around place to place, that's all. He must have made one lakh[33] speeches around the state...well, at least twenty-five thousand.

Author: Would he just show up? Were there any advance organizers as you have now?

Pramodh: Not at the beginning. He himself went, and just collected people under a big tree, or at someone's house. Very early meetings—in the mid-[19]70s—there were just 25, 30, 40 people at a time. After our first agitation in 1977, then people started approaching him to say "Come to talk at our village, at our meeting." Also, then others started spreading the word. They picked up his idiom. In this second phase, the communication started being handled by the upper section of the cadres—those who are not totally illiterate, not totally uneducated—capable people who were eager to do something but didn't have political connections. They did the work of spreading the message. We perfected this method by the mid-[19]80s. Now he [Joshi] need not go, but he still does when occasion demands. In fact, he's probably still the only one who's really traveling throughout the whole year.

In all of these face-to-face encounters with Joshi, other central leaders, or local and regional organizers ("the upper section of the cadres") the meanings and objectives of the movement are authoritatively articulated from a central leadership perspective. The symbolic complex of language, images, and ideas that signifies the movement at the overt level is reinforced. Moreover, at most of the Sanghatana gatherings that I attended, these speeches or lectures by leadership were backed up by other authoritative forms of movement communication. The largest Sanghatana gatherings are sophisticated productions involving podiums and sound systems, tented seating areas for members of the press, banners and posters, and performances of songs or readings of poetry that celebrate the movement, its objectives, its ideology, and its shetkari participants.

Some of the songs or poems performed at these gatherings are "traditional" folk verses or well-known verses from the independence movement era. In the context of a Sanghatana gathering, the performance of these adds new layers of Sanghatana meaning onto existing form and content. Others are verses well known by participants that have been generated within the movement, or new compositions by a local performer, as in the case of the following excerpt from a song I recorded at a mass rally in 1997:[34]

Male vocalist

At the place of injustice, you have been given more than
    your share.

Oh, shetkari, how you have slept!

Think of our Sanghatana, think of our unity, and you shall
    not be scared.

Get up, oh shetkari, your day has come.

People, be united!

A subsequent verse at this same rally, sung by a female vocalist, intertwined the Sanghatana's demand for fair pricing with the rich imagery of the Varkari Sampradaya movement—and implied that Sharad Joshi or any other leaders speaking at the podium were comparable to the folk deity Vithoba:

Female vocalist

The shetkari has died paying back his debts,

He put his sweat into the land, but never received a fair
    rate in return.

Nobody thinks about his life…

Light a torch; wake up this village.

Until today, nobody has heard the shetkari's voice.

Nobody has heard his cries.

But the Master of Pandharpur[35] has come running…

Wake up!

Until today, the shetkari was ignorant…

Wake up!

The Master of Pandharpur has come in the form of a
    shetkari.

He has come to demand justice.

Most frequently, however, songs similar to the latter make references to "the shetkari king" (which could be construed as Shivaji or Bali), or very pointedly to Bali himself—normally combined with references to the movement and its objectives that make it easy for willing listeners to associate these figures of veneration with the Sanghatana leadership, most specifically, Sharad Joshi.

Verses such as these are not direct productions of central leaders—rather, they are initiated by local organizers and crafted by local performers—but they are performances of movement identity that reproduce the overarching movement idiom expressed at the central level. Moreover, the prominence of these performances and the size of the audiences they are able to reach are only possible because of the scale of the communication resources available to movement leaders.

The Sanghatana's communications resources are not limited to rallies and posters and banners. Sanghatana leaders also represent the movement through publications of many types. For example, before any major rally or agitation, Joshi and his associates prepare and distribute press releases to national, state, and local newspapers. With some of these papers, the Sanghatana has established a sympathetic relationship—ensuring that the movement's activities receive good coverage with representation consistent with the ideas of Sanghatana leaders. Some Sanghatana leaders contribute directly to newspapers and journals—particularly Sharad Joshi, who writes a number of newspaper columns in English and Marathi.[36] In addition, the Sanghatana publishes and widely distributes a fortnightly newsletter in Marathi—called *Shetkari Sanghatak*—that is specifically tailored for communicating the movement's philosophy, agenda, and upcoming activities to organizers and participants. At the time of my fieldwork, five thousand bound copies of this newsletter were being mailed to villages around the state every two weeks.[37] Soon after my field research, the Sanghatana began taking advantage of electronic media, maintaining a website, communicating with some of its organizers via email, and distributing PDF versions of the newsletter.

The Sanghatana has also published book-length treatises on the shetkari cause. Through its informal publishing arm, called *Shetkari Prakashan* (Shetkari Press), the Sanghatana has published books in Marathi on such wide-ranging subjects as the legacy of Shivaji, the thought of Mahatma Phule, the rural women's movement, the history of agriculture, and economic liberalization. Most of these have been written by Joshi, and are intended for both rural and urban audiences. As a tool for communicating ideas to its participant base, one of these books stands out. The text, a transcript of lectures by Joshi entitled *Shetkari Sanghatana Vicar ani Karyapadhdati* (Thought and Practice of the Shetkari Sanghatana, 1982), is essentially a training manual for village organizers and activists. First published by the Sanghatana in 1982, it was distributed widely and became an important resource as a primer on the Sanghatana's agenda, its rationale, and its strategies for promoting change. Many activists have told me stories about village reading collectives that were organized for group study of this book. In these study groups, gathered under the shade of a tree or by the light of a lantern, literate individuals took turns reading passages aloud which were then discussed by the entire group.

Constructing meaning through organizing and forms of action
Just as the central leadership of the Sanghatana has a disproportionate capacity to define the movement through communication resources, it also has substantial control over the contexts of organizing and concerted action in which participants experience the movement. Communication, organizing, and action are deeply connected. Every speech, song, poster, or publication, for example, is an invitation to participate. Every agitation is a tool for drawing in new participants, an opportunity for participants to demonstrate leadership, and a vehicle for reinforcing the movement's ideology and agenda through performances of unity, rural power, and shared interest in its demands.

Photo 6. Sanghatana participants stop vehicles on a major highway in Marathwada region as part of a day-long campaign in cotton-growing districts, in 1997. After informing occupants about the challenges faced by cotton-growers, and providing them with literature about the movement's demands, participants allowed drivers to continue through the blockade.

The Sanghatana's patterns of organizing and action reflect the movement's propositions about broad rural unity. For example, important leaders of the movement have been a diverse mix of men and women, Hindus and Muslims, brahmins, Marathas, kunbis and Dalits. These leaders reach out to activists and participants from all social communities, from any political affiliation, and of any rural economic position. Thus, when local organizers and participants come together in a planned action, they are often mixing with each other across social divides that would, in many other contexts, keep them apart. At one small road-blocking agitation I observed in Jalna district, for example, the forty-five participants identified themselves as members of four different castes and five different political parties (including several who declared themselves as loyalists of the Hindu chauvinist Shiv Sena party). Some of these men were large landholders, dressed in fancy bush shirts or starched, white *kurta-pyjama;* some were smallholders or laborers, dressed in soiled and threadbare *dhotis* and undershirts; and a few were village merchants who were disconnected from the land. All of them were working together in the agitation, and sharing tea side by side during breaks from the action.[38]

Photo 7. A sign at the entrance to a village in Marathwada declares the village as a "Bali Rajya Gav." The Shetkari Sanghatana logo is on the upper left.

This same "coming together" of differentiated subjects can be seen in participation at rallies, training camps, *karyakarinis,*[39] or village reading groups, as well as in Sanghatana agitations. The fact that participation in some of these activities often entails high risk (of personal harm or trouble with the law) further cements these relationships. This experience of solidarity not only serves to validate the movement's core theme of shetkari unity, it also creates a social context for participants to communicate across their social differences in an idiom appropriate to the experience of solidarity—the idiom of the Shetkari Sanghatana.

In addition to actualizing the proposition of unity and community, collective action also structures participant understandings of the terms of the struggle and reinforces the movement's central diagnoses of and prescriptions for change. For example, the Sanghatana's commitment to nonviolence underscores the movement's philosophy of *bridging* relationships rather than fractionalizing them.[40] Tactics and strategies also define the line of struggle. Common Sanghatana tactics such as blocking road and rail traffic *(rasta roko* and *rail roko)*, withholding crops from market, closing villages to loan agents and politicians *(gav bundi)*—and more colorful tactics such as pelting urban politicians with onions,[41] riding corrupt local officials out of village boundaries on the back of a donkey, or shutting off power to Mumbai—all assert the integrity and "self-rule" of the "Bali Rajya villages" of the rural periphery and etch a boundary between Bharat and India.

Thus, participation in gatherings or actions is more than just an expression of shared interest in the movement's demands—it is also a performance of the movement's worldview and the movement community. Moreover, it is a performance that is corporeal. It affirms the legitimacy of people's participation through their bodily commitment, while also training bodies in performances that become signifiers of movement identity. These bodily engagements—whether traveling by foot, tractor, or bus to attend a rally, wearing the Shetkari Sanghatana badge, confronting police and traffic in a rasta roko, or raising fists and shouting "Sharad Joshi *zindabaad"* in unison with dozens or thousands of other participants—help to firm one's individual commitment to the idea of—and the ideas *in*—the movement. They inscribe the movement on the individual, linking the participant physically to the performative context and contributing to a dialectic between thought and experience that is generative of movement community and collective identity.

Conclusions

In this chapter I have discussed several dimensions of the Shetkari

Sanghatana's interpretive construct as it is represented at the macro level of the movement. These dimensions comprise three key interrelated aspects: (1) the movement's economic agenda and objectives; (2) the collection of signs and meanings that the movement draws upon, and; (3) participants' mass, public engagement and actions. Together, these three elements correlate with the terrain of movement identity and ideology that is most commonly vocalized by leaders and participants—and which, in turn, is generally invoked by outside observers when they represent the Sanghatana or other social movements in terms of a homogenous, monolithic identity.

In chapter 1, I introduced the fallacy of solidarity that extends from a premise of progressive solidification. In this chapter, we have explored the appearance and experience of solidarity in a more nuanced way. In order to understand the appearance of identity solidification, I first examined the applicability of the Sanghatana's agenda for higher prices in relation to the "on the ground" reality of the rural economic experience in Maharashtra. We considered numerous generalized subject positions and contexts in which the issue of pricing may be experienced by participants, and assessed the significance of these in relation to the Sanghatana's participatory base. What we have seen is that although rural economic life is experienced in widely variant ways, the Sanghatana's central issue of government control over agricultural remuneration does have the capacity to resonate, in at least situational ways, across a large cross section of rural Maharashtrian society. Moreover, the issue can be seen as metaphoric of numerous other policies of state control on agriculture that affect remunerative potential in ways that include, but are not limited to, controls on prices.

Next, I examined the Sanghatana's non-economic cultural idiom—a broad range of signs drawn from the Maharashtrian experience that includes, but is certainly not limited to, the ideal kingship of Bali. This idiom constitutes a second set of guiding ideas through which the shetkari plight is defined, its causes diagnosed, and its solution prescribed—but in terms that have rich cultural depth. As we have seen, the Sanghatana's popular cultural idiom serves several functions. One important function is that it helps to establish deeply meaningful points of agreement across rural identity groups that enable participants to effectively interact within the movement community on a plane of abstraction that at least partially or contextually supersedes other identities. A second, but equally important, function is that, by constructing an interpretive schema that directly accommodates or contests the signs and meanings embodied in other constructions of identity in Maharashtra, this popular idiom enables the Sanghatana to compete for the rights of signification and to insert its own meanings into a statewide dialogue on the key factors of Maharashtrian life. Part of the way it does this is by utilizing cultural resources to craft a narrative of inclusivity, continuity, and differentiation.

However, when considering the economic agenda and the popular cultural idiom as separate aspects, it is important to keep in mind that they are not discrete. A movement's stated agenda may be the most concrete and overtly purposeful component of its ideology, but it is also just one component of a much larger constellation of signs and meanings. For this reason, I have also examined key strategic resources through which movement leaders seek to maximize alignment between people's perceptions and experiences of the movement with the movement's diagnoses and prescriptions for change. What we have seen is that, although participant interests and subjectivities play an important role in the production of the movement and its meaning, movement leaders do have unique capacities because of their access to communication resources, their management of movement organizing, and their influence on the participant experience of collective action.

All of these interrelated aspects of the movement—its economic agenda, the collection of signs and meanings, and the experience of action and interaction—are conducive to the creation and performance of a movement identity and a distinctive idiom of identity through which participants interact and represent the movement.

Notes

[1] A shopkeeper. *Walla* is a common Marathi and Hindi suffix that refers to a person or thing that is, has, does, or is in some other significant way associated with the preceding word.

[2] Government of Maharashtra Department of Agriculture, 1993. Official report on the 1990–1991 Government of Maharashtra Agricultural Census. This was the most current statewide agricultural holding data that had been published at the time of my fieldwork and analysis.

[3] For simplicity, I have converted all land measures from metric or regional conventions to acres. In this case, the corresponding census figures in hectares are: Marginal Holdings: up to 1 hectare; Small Holdings: 1–2 hectares; Semi-Medium Holdings: 2–4 hectares; Medium Holdings: 4–10 hectares; Large Holdings: more than 10 hectares.

[4] Held in the city of Akola, in December of that year. My own estimate of the total number of participants (calculated on the basis of the "four square foot" rule mentioned earlier) was 30,000 attendees.

[5] For assistance in this survey, my thanks to Manjusha J. and Sambhaji P.

[6] Government of Maharashtra Department of Agriculture, 1993.

[7] In my research I have used surveys as supplemental instruments to the more conventional anthropological toolkit of ethnographic interviews and participant observation. Although surveys enabled access to far greater numbers of participants than would have been possible through other techniques alone, I have not depended on them for pivotal data in my analysis. One problem with over-reliance on surveys is that they are prestructured according to the researcher's own assumptions of what is important—or, as Spradley (1979) puts it, "they begin with questions rather than a search for questions" (32). Furthermore, as an extremely obtrusive method of inquiry, highly structured surveys run the risk of eliciting responses that are staged (so that the response is tailored to how the respondent wishes to represent himself/herself in front of an unknown inquisitor) or meet only the most minimal perceived expectations for information sharing. Another reason for avoiding over reliance on surveys is that, in the context of the largest types of gatherings described above, it is difficult to ensure rigidly scientific sampling techniques and sample sizes. This, however, was not an issue at small gatherings (such as the road blocking agitation in Aurangabad), where my sample was 100 percent of the population of participants.

[8] The most prominent movements in India that have been identified and scrutinized as "new farmers' movements" (NFMs) are, in addition to the Shetkari Sanghatana, the Bharatiya Kisan Union (BKU) in UP, under the leadership of Mahender Singh Tikait, and the Karnatak Rajya Ryot Sangha (KRRS) in Karnataka, under the leadership of Professor Nanjundaswamy (see contributions to Brass, 1994b, 2000; Byres, 1994). These movements (as has been the case with Kisan Sabhas and conventional Left movements) have adopted some of Joshi's concern for pricing, but have predominantly stayed within the conventional demands of subsidies and state-sponsored redistribution. The member movements of the Sanghatana-affiliated ICC (including the BKU in Punjab, the BKU in Haryana, the Khejdoot Samaj in Gujarat, and some smaller affiliates in UP and MP) are minor in size and influence compared to the Sanghatana, BKU, or KRRS.

[9] As of April 1997, these zonal restrictions on cane sales were slated for de-legislation. The Sanghatana was a leader in the fight for this, primarily because it was a restriction on producers' freedom to sell to the bidder of their own choosing—but also because it would help to reverse the politicization of the cooperative sugar factories. One of the reasons that the Sanghatana was able to achieve a victory in this campaign is that the state's governing Shiv Sena party also saw this as an opportunity to fracture the last vestiges of Congress Party power in the Desh, which was largely dependent upon the cooperatives for political loyalty.

[10] The "license-permit Raj" refers to a highly bureaucratic system under which wide ranging forms of economic participation are regulated by government oversight agencies empowered to grant, extend, or withdraw business licenses. These regulations, which were largely introduced under British rule and continued to expand under Congress Party rule, are widely regarded as a barrier to free enterprise and an entrenched mechanism for corruption and cronyism.

[11] Though this quote is nearly always paraphrased in one way or another by leaders and activists, I often heard it phrased as "the only thing the poor need is for the *government* to get off their backs." This is a notable addition to Gandhi's own words, in which he was less clear about the actual positioning of the oppressor: "I know village economics. I tell you that pressure from the top crushes those at the bottom. All that is necessary is to get off their backs" (quoted in Kishwar 1992, 7).

[12] Charlesworth, 1972.

[13] "Negative subsidy" is nearly always used within the movement as an English language term, regardless of the language in which it is written or spoken.

[14] These figures are widely used in public speeches by Joshi. Also see Kishwar, 1992. Joshi is equally opposed to positive subsidies as he is to negative subsidies. He uses these figures to dramatize the disparity between the two, but advocates a subsidy-free and competitive global market economy.

[15] Compared with land data, which is maintained by village panchayats and/or *patwari* record keepers, accurate estimates of the population of landless agricultural labor in Maharashtra are even more fraught with difficulties. In practice, the number of fully landless agricultural laborers is difficult to determine for a number of reasons, including: 1) estimates may include marginal landholders who rely on wages from agricultural labor to make ends meet; 2) many laborers are migratory and present challenges for census takers; 3) many people who appear as agricultural laborers are seasonally employed in rural areas, while spending off-peak periods in the informal or semiformal sectors of the economy in urban areas; 4) it is not uncommon for laborers to have some unrecognized access to plots—either land of undeclared ownership or plots that have been squatted on public land. These are most likely to go unrecorded.

[16] This and the factors given above indeed may be some of the reasons that laborers appear to be relatively under-represented in my own surveys.

[17] Carter (1988), in his excellent discussion of household dynamics as a factor of observable changes in aggregate land distribution patterns, makes it clear that the available data raises more unknowns than knowns as to who is acquiring or selling what kind of land and for what reasons. Nonetheless, he does see an overall pattern in Maharashtra in which "on average, and in nearly all communities, landless households and those with small holdings have purchased more land than they have sold and they have done so at the expense of larger holders" (151–152). I should be clear here, however, that I do not intend to overstate this case or to imply that this should be understood as an egalitarian trend in Maharashtra. As Carter points out, one explanation for this redistributive trend could be that larger landholders have traded-in large parcels of dry, unimproved and unproductive land for smaller but vastly higher yielding irrigated land—selling the worst land to laborers and marginal landholders at relatively affordable rates.

[18] While Varshney is particularly writing about the Indian context, the import of this insight should not be restricted to India. As a wide range of recent social movements, political uprisings, and nation building experiments around the world have shown in recent decades, India is by no means extraordinary or unusual in the extent to which ostensibly noneconomic identities underlie social mobilization.

[19] Bharat—a term that is both Hindi and Marathi—is not novel to the Sanghatana. As a common term for "India" in any vernacular speech context, it is a word of extremely broad signification that has been used with great variances in meaning and objectives by wide ranging political interests. Moreover, its use as a signifier of "rural" India to the exclusion of urban and governmental India is also not new. This traces to the once extremely popular kisan movement of Chaudhary Charan Singh, who left the Congress Party in Uttar Pradesh state in 1967 to found a rural-interest opposition party called the Bharatiya Kranti Dal (Bharat Revolution Force) or BKD (Gupta 1998).

[20] This and the following quote are from a transcript of Joshi's lectures at this training camp. S. C. Mhatre, who later compiled Joshi's lectures at this camp into what has since become the Sanghatana's essential activist training manual, indicated in his introduction that the trainees were "all from rural areas and were little educated." "Naturally," he goes on to explain, "the talks and discussions in the workshop about the ideology and tactics of the Shetkari Sanghatana were conducted in simple language." All translations from this text are a joint effort by myself and Arati K., who assisted me in numerous transcription efforts.

[21] The Tata and Birla families are two of India's largest and most successful industrial houses. The Tata collection of industries includes textiles, chemicals, food products, hotels and tourism, hydroelectric power, steel, locomotives and automotives, publishing, electronic information technology, management consultancy, and financial services. In addition to India, the Tatas hold concerns in Africa, Europe, the Americas, and in other parts of Asia. These names are well known to villagers not only because they use household products and agricultural implements that bear these names, but also because the names are often used as shorthand for power and prestige—much as one might use the name Rockefeller in the United States.

[22] The adjectival construction of Bharat.

[23] This occurred in November 1983.

[24] Jowar (sorghum) is a staple coarse grain in rural Maharashtra, and is not a "rich farmer" crop. As a food crop better suited to Maharashtrian dry land conditions than other more expensive grains like wheat, it is widely grown by marginal holders who rely on it as a low risk cropping strategy. If the harvest is good, surplus may be taken to market; if the harvest is bad, the entire crop may be retained for household use.

[25] This poster was part of the organizing campaign for a gathering of rural women to be held at Chandwad, Nashik District, on November 9–10, 1986. About 300,000 women attended the gathering, which was designed to give attendees a chance to deliberate on issues specifically affecting rural women.

[26] This poster and the following one were both prepared for the Maharashtra State Assembly election of February 1990. In this election, the Sanghatana promoted the candidacies of several Sanghatana participants in association with (and under the election symbol of) the Janata Dal party headed by Mr. V. P. Singh.

[27] Vamana's name appears in quotation marks on this poster, in order to highlight a clever play on words. Since Vamana is the name of a dwarf, in popular usage this word can also mean "unevolved" or "half-baked." Thus, the phrase "communalist Vamana" can also mean something like "the half-baked ideology of communalism." Such double entendres in wording are common in Sanghatana slogans, and reveal some of the playfulness and versatility of cultural language and signs that is not easily achieved through strictly economic language.

[28] As it has been conceptualized in the literature, the framing of a movement is largely a top-down affair (Oliver and Johnston 2000). This contrasts strikingly with recent scholarship emphasizing that movement cultures are dynamic and participatory creations rather than simply the constructs of movement leaders (cf. Poletta and Jasper, 2001; Robnett, 1997; Smith, 1991). It is important, therefore, that we do not confuse leadership strategies to construct a collective action frame with the generation of all movement culture or movement identity in general. However, we must also recognize that mass movement leaders have unequal access to tools of signification and that they have a vested interest in the power of signification to mobilize diverse subjects. With this in mind, sociologists Pam Oliver and Hank Johnston (2000) have argued that the concept of framing works best when we use it not to understand the overall production of meaning within a movement but

rather to make sense of the specific strategies and resources that leaders employ within a movement-wide competition for ideas.

[29] Here Joshi uses "Hindustan," another common term for the country of India. In this case, Joshi is using this third term to describe an entity that encompasses both "India" and "Bharat." In most contexts, Joshi refrains from frequent use of this word because—due to its perceived (but historically inaccurate) affinity with "Hindu" religious practice—it has become the preferred term of the Hindu nationalists.

[30] Swarajya—along with *swatantra* (independence)—was a central rallying call of the independence movement. Although surajya was more or less implied as a natural outcome of these, it was not as commonly articulated in the independence movement. Joshi's point here is that the one does not guarantee the other.

[31] Joshi, an urban brahmin from Pune, worked for several years with the International Postal Union, during which time he was based in Switzerland. Unlike other agrarian leaders in India, Joshi makes no pretenses about being a "son of the rural soil." For example, when he addresses rural audiences, he generally shows up wearing jeans, tennis shoes, and a golf shirt rather than conventional rural garb.

[32] This informant was speaking in English. Here "farmer" is his own word choice.

[33] Lakh = one hundred thousand.

[34] This mass rally was held at Kaundinyapur, in Amaravati district.

[35] "Master of Pandharpur" is a common reference to Lord Vithoba.

[36] Many of Joshi's columns are high profile and tailored for urban audiences, such as a weekly column on economic policy that he wrote in English for the *Times of India* called "Bharat Speaks." During a visit by the author with Joshi in April 1998, he enumerated his recurring column-writing engagements:

> *Shetkari Sanghatak* [the Sanghatana's Marathi fortnightly newsletter]—every two weeks; *Deshonatti* [a Marathi daily with regionally targeted editions]—writing a column every two weeks on what I consider to be pivotal books in economics; *Times of India* [a nationally circulated English language daily]—contributing "Bharat speaks" every two weeks; *Business India* [an English language weekly journal]—my column, every two weeks; and *The Hindu* [a nationally circulated English language daily]—every two weeks, starting soon. The *Business Standard*

also wants me to start doing a fortnightly column. I'm still thinking about it.

[37] This was the fortnightly print run on the newsletter when I spoke with its editor in 1999. At that time, recipients paid an annual subscription fee of 30 rupees.

[38] Organizing and action in the Sanghatana is, in this sense, markedly different from most other contemporary Indian agrarian movements. The BKU (Bharatiya Kisan Union) movement in Uttar Pradesh, for example, organizes primarily through its leader Mahender Singh Tikait's Jat caste social structure, which is closely aligned with Tikait's own political party affiliation. Moreover, Tikait and the BKU define "kisan" as a landed cultivator, and thus do not actively seek participation from laborers or other non-cultivating members of the rural community (see Gupta 1997).

[39] Annual constituent assemblies where issues and strategies are discussed and decisions made for the year ahead.

[40] Dhanagare (1994) raises an interesting perspective on this, suggesting that the nonviolent character of the Shetkari Sanghatana supports his thesis that it is a rich farmer project. Rich landowners, he suggests, have more to lose than to gain by "violent mass upsurge and destructive agitations," which not only destroy property but upset the social order as well. While there is probably something to this, it overlooks the fact that the Sanghatana is not a revolutionary struggle determined to capture or disable infrastructure and resources. On the contrary, the movement's greatest interest lies in being perceived as an extremely nettlesome opponent, but one that is also a reasonable and trustworthy negotiating partner.

[41] In June 1997 the Sanghatana launched an "onion pelting" agitation aimed at drawing attention to the plummeting prices that growers were receiving for onions, and at humiliating political figures who were not responsive to the Sanghatana's demands. In Nashik district, this led to a number of arrests in different parts of the state, when participants threw onions at the cars of government ministers (*IE*, 6/24/97, 2). The agitation eventually spread all the way to the halls of state legislative assembly, causing a great commotion when legislators of the opposition (who were sympathetic to the onion growers' situation) rose from their seats and pelted onions at the members' benches of the state treasury department.

4—Dialogics of interest and identity

This is a single issue movement: the issue of price. This means the issue of liberalization, of liberation for our women, of freedom to have a life without poverty and debt. This is what we mean by Bali Raj.

—A local Shetkari Sanghatana Women's Front organizer, Kaundinyapur, Maharashtra

During my field research on the Shetkari Sanghatana, I met many people who had been deeply engaged with the movement for years on end. I met many others who had only recently become engaged or had been deeply engaged in the past but had chosen to largely or completely sever their ties, moving on to other vehicles for the pursuit of their ideals and interests. Most commonly, I met people who drift in and out of the movement on a recurring basis, and who profess allegiance to other arenas of opportunity and other terrains of identity in other contexts. At first, saddled with presuppositions that I would find a Shetkari Sanghatana form of engagement and identity that was consistent and relatively homogenous, I felt unsure that I had managed to find the "real" sanghataks who were the true participatory force of the movement. After several months, however, I realized that this was them.

What I came to understand is that for most participants the Sanghatana represents just one of many different opportunity structures in which they may be engaged. Some of these structures are formal—other movements, political parties, government- or NGO-sponsored village development schemes— and others are informal social, economic, and political relationships that may be bolstered by Sanghatana participation, threatened by it, or neither. Many of these structures represent constellations of objectives and meanings that are relatively contiguous and overlapping, while others may be far more distinct, or even opposed. Thus, participants often shift their performances of community and identity from context to context in ways that may be either very subtle or readily discernable.

In this chapter and the next I will examine the Shetkari Sanghatana as a negotiated space, crafted not simply according to the designs of movement leaders but through participatory shifts and dialogues that involve a wide assortment of participants and interests. First, in this chapter, I will consider the contextually shifting nature of participation in the movement and how shifts in participation can help us dig beneath the surface of the Sanghatana's representation as a single-issue movement. Then I will look more closely at the dialogical construction of other important issues and objectives that motivate people to participate, and consider ways that the Sanghatana's organizational structure and practices actually facilitate, rather than inhibit, broad dialogue on which issues and interests the movement should serve. In the next chapter, I will look more closely at the construction of deep cultural meaning in the movement from a similar perspective, focusing particularly on the idiom of Bali Raja and the restoration of the demon king's reign.

Peeling the onion of the single-issue movement

The Shetkari Sanghatana has been immensely successful at recruiting participants from a broad cross section of rural society in much of Maharashtra, but it has not been equally successful in every part of the state. Maharashtra encompasses widely divergent social histories, variable access to resources and political patronage, and assorted regional and community identities. Thus, while the movement has been strong in Marathwada, Vidarbha, and parts of Western Maharashtra, in other parts of the state its expansion and organizing successes have been limited by alternative structures of political patronage, other dominant forms of power, and other dominant constructions of self- and community-interest in economic life.

Clearly, despite their distinctive resources, the Sanghatana's leaders cannot simply fabricate the movement at their will, deciding *where* it will thrive and *who* will participate. While this should seem self-evident, it is a point sometimes lost on observers who portray the Sanghatana as a "rich farmer" movement, and endow it with an overwhelming hegemonic capacity to recruit mass participants whose interests it does not serve.

Just as there are perceivable limits to the power of its elites to craft the movement in those areas where it has seen its least success, these limits also apply in Vidarbha, Marathwada, and parts of Western Maharashtra where the movement has been its most successful. Even in these very successfully organized areas, the demand structures and representation of interests in the movement are as diverse as the regions' environmental and social landscapes. Demands and representation are informed by the differing economic and social realities in each district, within each village, and within different segments of village communities.

As I have argued, there are indeed contexts of experience in which a broad cross section of rural Maharashtrians in these regions may find rational, interest-based common ground in a struggle for higher prices. This common ground, however, may only occur in some contexts of action—particularly for subjects who only occasionally identify with the interests of market oriented cash-croppers, opting at other times to identify with labor interests, caste interests, political party affiliations, and other terrains of action and identity through which their individual interests may be best advanced. Price, therefore, can be a functional motivator of participation in some contexts, but not in others. Accordingly, we should expect to find that individual participation in the Sanghatana, rather than remaining constant, would ebb and flow over time, shifting from context to context. Just as importantly, we should also expect to find that the Sanghatana, as an organization seeking to broaden its following to include the widest possible range of rural inhabitants, would make efforts to address many other issues relevant to its actual and potential base.

Indeed, when we look at the movement in terms of the broader calculations of interest that underlie the participation of different people at different points in time, a picture of the Sanghatana's action agenda emerges that is much more expansive than what central leaders themselves refer to as the One Point Plan.

When one looks more closely, the Shetkari Sanghatana appears less like a single-issue movement and more like a multi-tiered collection of allied movements that undertakes a range of activities on a wide variety of issues. Many of these activities represent other types of demands and prescriptions for change that are only tangentially, if at all, related to remunerative prices. For example, the Sanghatana is engaged in a number of statewide political reform initiatives, involving rural political training, corruption eradication, and autonomy issues. Three of its major initiatives in this realm are:

Launching an alternative political platform for rural leaders: In 1993 the Sanghatana founded the *Swatantra Bharat Paksha* (Independent Bharat Party) (SBP). The SBP is an unofficial, unregistered party loosely affiliated with the Sanghatana, designed for promoting the political candidacies of rural individuals sympathetic to rural problems.[1] Through its campaign committees in individual districts, the SBP has helped to build rural leadership capacity in the formal political arena, and it offers an alternative platform for political representation in villages around the state.

Monitoring and exposing corruption: In 1997, the Sanghatana launched a statewide anticorruption program called the "Q movement." Under this program—which stands for "quit corruption," a slogan reminiscent of the Quit India movement of the colonial era—local Sanghatana groups establish public Q *kendras* (Q centers) in their villages where all members of the community—whether Sanghatana participants or not—can register cases of corruption or harassment by state, regional, or village authorities.

Demands for regional autonomy: These entail, specifically, the Sanghatana's demands for the separate regional statehood of Marathwada and Vidarbha. As discussed earlier, the prospect of statehood for these regions is attractive to many local agriculturalists, who feel that the surplus value of their production is skimmed off by the state in order to fuel industrial and agricultural development in the regions to the west. The statehood movement, however, is also appealing to many low-caste, Dalit, and Muslim residents of Marathwada and Vidarbha, who feel that their populations (which are proportionally high in these regions) are currently underrepresented in statewide politics.

The Shetkari Sanghatana also embodies a wide variety of highly organized, movement-wide initiatives focused on rural women. These include:

Women-led alternative agriculture: In a statewide program called the *Sita-Sheti* campaign, Sanghatana activists train women in cropping and gardening practices that use the smallest amount of expensive inputs such as pesticides and chemical fertilizers.[2] This not only helps small plot holders transition to lower cost food production methods (whether they are participating in the market or producing only for household consumption), but also creates opportunities for women who raise garden produce to participate in the increasingly lucrative urban market for organic and chemical-free foods.

Registering women as property owners: In 1990, the Sanghatana launched a campaign called *Lakshmi-Mukti*[3] in which male agriculturalists are encouraged to voluntarily transfer part of their land title to their wives. This initiative is aimed at insuring a woman's inheritance of property in the event of her husband's death. The program has been remarkably successful. Within a year and a half of its launch, more than 100,000 women benefited from land transfers that were legally facilitated by the Sanghatana.

Promoting women's representation in village *panchayats*: Across the state, Sanghatana leaders have spearheaded a movement to get village women elected to their local panchayats (village governing councils). The aims of this initiative are to help village women develop leadership skills and to create local councils that are attentive to the welfare of women and families in the issues they undertake and in their allocation of village development funds. In many areas the movement has succeeded in establishing panchayats that are 100 percent female.[4]

In addition, there are a number of initiatives that are focused on marketing agricultural products and protecting the value of agricultural land. These initiatives, which could benefit large landholders and small-holders alike, include:

Cooperative marketing schemes: In a number of villages, I have spoken with groups of sanghataks who have established cooperative marketing schemes and shops. These enable agriculturalists to sell certain types of produce direct to market, cutting out the middleman and redistributing profits among participants based on the amount of produce that each deposits into the co-op.

Land development companies: In several parts of the state, Shetkari Sanghatana groups have experimented with organizing both small and large landholders into joint stock companies. These have become platforms for landholders to leverage compensation demands in situations where their land is at risk of being taken over by industry or government for public and private projects such as surface irrigation schemes, industrial complexes, and highway or airport expansions. I will discuss an example of this in more detail below.

Finally, there are also a wide range of demand and non-demand initiatives taken up by the movement at the state, regional, or village level that appeal to the interests of a very wide range of rural inhabitants, irrespective of their position in the rural economy. These include:

Infrastructural demands: These are often very localized demands, undertaken by individual Sanghatana chapters. They include demands for new roads that connect villages and hamlets to the outside world, or demands for improvements in local water supplies, electrical supplies, and schools. The benefits of such initiatives extend very broadly across a village population.

Compensatory demands: These include demands of government compensation for agriculturalists who have lost crops due to drought or adverse rains. They also include demands of debt forgiveness for struggling villagers whose electrical bills, bank loans, or other accounts are in arrears. Such demands are often taken up locally, but have also been the focus of numerous movement-wide campaigns.

Temperance initiatives: In some villages where women have complained about spousal abuse and the squandering of scarce household finances due to their husbands' alcohol consumption, local Sanghatana participants have coordinated programs for promoting sobriety. In many locations, women activists have succeeded in shutting down the operations of local liquor purveyors.

Labor and Dalit rights initiatives: In many parts of the state, Sanghatana leaders have at different times partnered with laborers' unions and Dalit organizations to promote human rights and encourage landowners to pay voluntary wage increases. These initiatives are embedded in many other aspects of the movement. One of the terms under which a village may declare itself a *Bali Rajya gav* (a Bali Rajya village), for example, is an agreement from all labor-employing landowners to pay laborers a fair wage that exceeds the government-determined agricultural minimum wage.

These are just a few of the many different issues and programs that have been undertaken at different levels of the movement. Quite aside from the narrow view of the Sanghatana based solely on its representation as a single-issue movement with a One Point Plan, this range of activities reveals that the Sanghatana has many different but interrelated agendas. Many of these are not even in the form of demands placed before the government, but are, rather, grassroots initiatives aimed at promoting change and developing resources for change within the village itself. Each of these articulates new issues of relevance to actual and potential participants, establishes new opportunity structures through which participants may pursue their interests, and generates novel forms of organizing and collective action that further actualize the Sanghatana community through the experience of participation.[5]

In one way or another, most of these initiatives can be ideologically tied back to the *spirit* of the One Point Plan— particularly in its broadest sense as a metaphor for outside control. But these specific issues appeal to a much greater cross section of rural society, and to many more contexts of self-interest, than the One Point Plan for remunerative pricing ever could on its own. When we look more closely at the movement on the ground, it should not be surprising to find a much broader range of participants and much more varied forms of participation than we might expect.

Distributed objectives and shifting participation

If issues and opportunities within the Shetkari Sanghatana evolve over time, then it makes sense that participation within the movement is diverse and changeable in response to the range of interests and opportunities it represents in any particular context. What then does it mean when an individual declares himself or herself to be a Shetkari sanghatak? What does it mean when an entire village declares itself a Bali Rajya gav? The implication, of course, is that these individuals or villages are fully engaged in the movement and have established their unfaltering allegiance.

In most cases, however, the reality is quite different. The movement is dynamic; participants and villages join the movement and, very often, participants and villages leave. People might enthusiastically participate for a time when an issue undertaken by the Sanghatana is of particular importance to them—but their participation may just as quickly wane after an issue has been successfully resolved. Moreover, participants may encounter new platforms for extending their interests or new impediments to participation that cause them to alter the depth or quality of their engagement with the Sanghatana.

Two mobilized villages in the Shetkari Sanghatana[6] offer an entry to understanding how the diversity of issues and opportunities can drive different forms and degrees of participation. In the first case—a village in Western Maharashtra that I will call Perugaon—the issues that motivated wide participation were demands and initiatives that had nothing to do with crop prices, despite the fact that agriculturalists there produce of one of the region's most important market crops. The second case, a village that I will call Dorlapur in Marathwada, offers insight into varied motivations for participation across a village society—some villagers expressing primary interest in the price-based struggle, but others identifying different interests to pursue through the movement.

Villages come and villages go—the case of Perugaon
Perugaon is in a hilly zone of the Desh subregion of Western Maharashtra. It is a village of less than 100 households, almost all of which identify themselves as either Maratha or Mali by caste. In some contexts, informants describe the village as entirely Maratha, and express deep pride in their Maratha heritage. Informants also describe Perugaon as a village that is tightly unified politically and socially. The village *sarpanch* (the head of the panchayat council) is supported and sanctioned by the village's most prominent family of hereditary (but now unofficial) headmen. Although this suggests a definite hierarchy of power within the community, land distribution is relatively egalitarian and the village has seen few disputes. Every family in the village

can claim ownership to some amount of land, but no nuclear family in Perugaon has direct ownership of more than five acres.[7] Thus, as defined by the Indian Census, everyone in the village is a smallholder or below.

As with many other villages of Western Maharashtra, Perugaon lies within a rain shadow area that is prone to both recurrent droughts and unpredictable heavy rainfalls. As such, agricultural livelihood is often tenuous, particularly considering the relatively small average plot size cultivated by village families. Unlike much of the rest of the Desh subregion, Perugaon has not benefited from state-sponsored surface irrigation schemes or the establishment of deep borewells. As a result, water supplies are not sufficient for the cultivation of sugarcane, the most lucrative crop in some neighboring districts. Instead, most cultivators in the village depend heavily on onion as their principle cash crop, and they also grow some coarse grains, peanut, chili, potato, and occasional garden vegetables such as tomato. One advantage that families in Perugaon *do* have is the village's proximity to a fairly large and industrializing market town, less than ten kilometers away. This enables speedy and inexpensive transport of crops directly to buyers in town, as well as seasonal or supplemental employment opportunities in town that would not be practicable from a greater distance.

In the late 1990s, Perugaon joined the Shetkari Sanghatana for the second time, after more than a dozen years of keeping the movement at arm's length. The first time the village was energized by the movement was in the early 1980s, when Sanghatana activists from the region made a new proposal that directly affected Perugaon: the activists proposed putting pressure on the government to improve the main road that linked all the villages of the area to the market center. Up to that time, villagers had been challenged by the deteriorating quality of the heavily rutted and unpaved road, which caused expensive damage to the wooden wheels of their bullock carts and, during the monsoon rains, was often impassible. Perugaon joined with the activists and embraced the Sanghatana. Villagers painted

Sanghatana slogans on village walls, they joined with residents of other neighboring villages in a rasta roko agitation that blocked traffic near the market town, and they risked confrontations and *lathi* (baton) charges with the police. Eventually, their pressure tactics worked; they won the government's promise (and ultimate reality) of a paved roadway. Shortly thereafter, nearly everyone in Perugaon broke off their formal involvement with the Sanghatana for well more than a decade.

During this gap between the early 1980s and the late 1990s, no one could have called Perugaon a "Shetkari Sanghatana" village. Although Sanghatana activists approached villagers with new issues that might have served their interests, few of them were willing to attend gatherings or get involved. And although its residents were growers of onion, the Sanghatana could not even convince Perugaon to participate in crop-price campaigns—even though the movement had established its reputation—with the first major agitation of 1980—as an ally of onion growers with the power to win higher prices from the government.

What happened after the victorious campaign for the new road is that most people in the village no longer saw an immediate benefit to participating in the movement. Building a new road was a very local demand that required their support in order to succeed, but (in a classic example of the "free rider" problem) most villagers assumed that the more general statewide struggle for prices would carry on without them. Their proximity (and improved access) to the market town meant that many villagers had well-established relationships with buyers of produce— private buyers, as well as agents of government purchase schemes—and they saw no reason to personally rock the boat. Perhaps most importantly, the movement lost the support of the village's leading families and elected leaders, who no longer perceived any advantage in aligning themselves with the Sanghatana. These families saw a conflict between alliance with the movement, on the one hand, and their relationships with important figures in formal party politics, on the other. Hence, for the rest of the inhabitants in the tight-knit village, enthusiastic participation with the Sanghatana would have threatened their social capital and economic relationships with power right at home.[8]

By sometime in the late 1990s this calculation of interests and opportunities had changed. By this time the nearby market town had become a locus of significant industrial growth, and developers had set their sights on building new industrial estates further down the road toward Perugaon. Under the authority of the government-run Maharashtra Industrial Development Corporation (MIDC), several thousand hectares of land from the area had been slated for acquisition by the state under public domain laws, affecting Perugaon and a number of other villages. Nobody knew exactly whose plots would be acquired, and all of Perugaon's landowners were concerned. Being forced to relinquish their land would be bad enough—but to make matters worse, the acquisition plan established MIDC as the sole monopoly buyer, effectively enabling it to set its own price for the buyout without having to compete with any other bidder.[9] The effect, as one young woman in Perugaon named "Sunita" described it, would be that MIDC would "steal the land from the shetkaris and sell it to industry." Elaborating on this, she added: "It would be exactly what happened to Bali. India will come and take whatever it will, and Bharat will get nothing."

Sharad Joshi and local Sanghatana activists got involved. With relatively little effort, they were able to convince the residents of Perugaon to join in a concerted effort to resist the monopoly buyout. Within months, every family in Perugaon had registered their land in a joint stock company set up by Sanghatana leaders—each landholder receiving certificates of shares based on the market value of their land—and the village joined in a Sanghatana initiative to put pressure on the government to eliminate MIDC as the middleman. Moreover, all of the landowners pledged never to sell if they were approached by MIDC or by any other buyer independently, regardless of the price offered. If developers wanted to buy land, they would now have to buy directly from all the unified shareholders of the shetkaris' new company. Thus, fifteen years after walking away from participation with the Sanghatana, the residents of Perugaon were once again wearing Sanghatana badges, painting Sanghatana slogans on the village walls, describing their struggle as a step toward the realization of Bali Raj, and raising their fists with cheers when Sharad Joshi took to the platform.

The example of Perugaon suggests that participants themselves—and not movement leaders—play a pivotal role in deciding when and how the Sanghatana can help them to advance their interests, and in what contexts individuals are willing to express solidarity with the movement through the symbols and idioms of movement identity. It also demonstrates how ground-level participation functions as a component of a larger dialogue on the meaning and agenda of the movement. Without Perugaon's participation in either of the key initiatives in which it became involved, the Sanghatana would never have pursued these as formal demands that could be backed up with the pressure of popular agitations or commitments of solidarity.

Fractures and unity in the Bali Rajya village—the case of Dorlapur
The village of Dorlapur dramatizes the same key role of individual participants in determining their own interests and in helping to define the movement's agenda, but the Dorlapur case is somewhat different. While Perugaon exemplifies broad and enthusiastic participation on specific issues from which nearly everyone in the village could benefit, Dorlapur's collective participation stems from a more complicated confluence of dissimilar or only partially overlapping perceptions of interest and opportunity.

The self-declared Bali Rajya village of Dorlapur, located in a predominantly dryland area of Marathwada, is typical of many villages in the region and contrasts sharply with the situation in Perugaon. Similar to the latter, the village is effectively ruled by one family of hereditary patils, whose political blessing is a prerequisite for election to the council seats of the panchayat. Unlike Perugaon, however, Dorlapur is heavily stratified both socially and economically. Its principal family of influence controls access to one hundred and fifty acres of the village's best land, with rich soil suitable for the cash cropping of cotton. Moreover, several members of the family have college degrees and entrepreneurial interests in the district's largest town. Their financial position, control of village resources, political ties outside of the village, and cultural capital are overwhelming compared with the majority of the village residents.

The caste composition of Dorlapur is substantially more diverse than that of Perugaon. While the leading families identify themselves as *assal* (true) Marathas—genealogically linked to the great historical Maratha chieftains of the region—many other members of the community identify themselves with kunbi castes that might, in some other villages, have identified themselves as Maratha. In most contexts of Dorlapur's village life, the distinction between these groups is well understood and routinely reinforced. Dorlapur also includes a sizable community of Mahar Dalits, who represent about 15 percent of the village population. This segment of the community, as is typical of the region, resides in a caste-segregated *Maharwada* on one edge of the village, outside the now crumbling village walls. It is a neighborhood of about twenty-five dilapidated mud and stone homes, where livelihood is based on seasonal labor in the cotton fields and a range of unskilled odd jobs in and around the village. Although most of the Mahar families have access to some land, these are without exception small plots of generally poor quality, which were either acquired on their own, or given to a family's ancestors by the village chief as *balute* for services rendered to the village.[10] On these plots they are able to cultivate a small amount of produce, almost exclusively for home consumption.

The leading families and most village landholders in Dorlapur had been active in the Sanghatana for at least a decade. They participated in a statewide "symbolic" one-hour rasta roko, called by the Sanghatana in 1986 to protest the low procurement prices on cotton. A month later, they participated in a massive railroad blockade chiefly focused on the same issue. In the years that followed, they continued to strengthen their ties with the movement—eventually joining thousands of other villages in the declaration of Dorlapur as a Bali Rajya gav. When I first visited the village in 1997, it was at a time when the state government Cotton Monopoly Purchase Scheme had declined for the third year in a row to raise prices in line with cost of living increases. The Sanghatana was again demanding a response from the government on the cotton issue and stepping up its cotton-focused agitations. Dorlapur was becoming even more active than before.

During my visits to Dorlapur, most activists and cash-cropping participants in the movement described the participation of villagers in the Sanghatana as "total"—*everybody*, they said, was involved. This was puzzling. Unlike the case in Perugaon, where the issues that had galvanized the village seemed relevant to everyone's interests, it was not as easy to make sense of the participation of Dorlapur's large community of Mahars in a movement that, within the village at least, seemed overwhelmingly focused on the issue of cotton prices (a crop that requires much more land and resources than anyone in the Mahar community had access to). My conversations with the Dorlapur's larger landholders offered little insight. Most of the larger landholders vaguely professed that the village's Mahars participated because higher prices would eventually lead to higher wages for their labor. Some of the larger landholders also suggested that Mahar laborers felt an obligation to support the landowners under the terms of established patron-client relationships. This particular understanding of laborers' participation was clear, for example, in a conversation that I had with a large landholding cotton cultivator in Dorlapur who I will call Rameshrao. During this conversation, Rameshrao, active in organizing an upcoming rally in the region, explained his plans for ensuring a good turnout on the big day:

> **Rameshrao:** The main issue is cotton price—so all the landowners, the cotton growers, will come. Also, all the merchants...because they support what we are doing. Really, anyone who has a

vehicle or can afford the ST [the State Transport bus] fare to the rally. The only ones who would not necessarily come are the laborers and the Dalit section of the village...but we'll just take them in our trucks and carts.

Author: And they'll agree to come, just like that? Do they actually want to go?

Rameshrao: See, those who work in my land will definitely come, because I ask them to and because they know that their interests are tied to my own. We have a long relationship...I provide the roti[11], I know their children, we greet each other on Diwali.[12] Also, they will see something interesting.

Rameshrao's comments reminded me of something I had heard just a few weeks earlier talking with a non-Sanghatana organizer of a sugarcane laborers' association, in the Desh region of Western Maharashtra. We had been having a discussion on this very same topic—the presence of marginal cultivators and landless laborers at Shetkari Sanghatana rallies—and this man, who I will call Suresh, shared his theory on the matter:

Suresh: These are the people that we represent in our labor movement. Yes, you may occasionally see them at a Shetkari Sanghatana rally, but they are not really supporters of that movement. What happens is that some big landlord or village leader organizes a truck or a tractor cart, and tells his laborers to get in. Then he takes them to the rally. It's an old political trick, intended to make a party or a movement look better, more important, through sheer numbers of people. But the people get in the truck not out of political interest, but at most because it is an adventure to take a trip...to see something new...and maybe

to be treated to a meal. And also, because they
have been told to come.

Suresh's perspective is a familiar one. Allegations of this sort
are common in Maharashtrian and Indian politics—particularly
in the context of campaign rallies by major parties in state and
national elections. In many cases, these allegations probably
embody greater or lesser degrees of insight into the actual forces
behind mass attendance. In the case of Dorlapur, however, I
quickly became convinced that both Rameshrao's and Suresh's
explanations did not fully account for the expected attendance of
the village's laborers and Dalits. In fact, their explanations
contrasted markedly from the analyses offered by laborers and
Dalits themselves.

The first time that I ventured into Dorlapur's Maharwada, I
interviewed a thirty-two-year-old Mahar laborer who we can call
Balu. This man had attended a handful of smaller Sanghatana
gatherings in the past, but had also been involved in Dalit-interest
meetings and agitations coordinated by another organization
called the Campaign for Human Rights (CHR).[13] Balu described
his intent to join with other villagers in the upcoming Sanghatana
event in the following way:

> **Balu**: It is true that what is good for him
> [referring to a specific labor-employing
> landowner] must eventually be good for us in
> some way. There is a relationship there. But the
> Shetkari Sanghatana can also be a relationship,
> just as with the CHR. If I participate in the
> Sanghatana, then, if there is some problem,
> someone else may help. If I am troubled
> unnecessarily or wronged by some official, I can
> record a complaint. If there is some injustice in
> the village, we can speak to other Shetkari
> Sanghatana people outside the village.

Speaking with Balu on another occasion, he described in detail his
feelings about Dorlapur's designation as a Bali Rajya village:

> **Balu**: Of course, this was not our idea [referring to the residents of the Maharwada], and nobody asked us if we should say that [Dorlapur] is now such a place...but we have always thought of this wada [the Maharwada] as the real place of Bali, and of ourselves as the real Bali Rajas. Who is it that toils? Is it that big *kapuswalla* [cotton grower][14]? Who's fathers and mothers were made to live here outside the village? So, when we are talking with those people we may say this whole village, Dorlapur, is a Bali Rajya gav...but when we do so we are really thinking of ourselves.

Balu's own description of interest highlights several opportunities that he perceives to extend from Sanghatana participation. They are different from the price objectives of village cotton growers, but equally valid. He acknowledged that attendance at the rally was a politically sound decision in terms of maintaining a functional relationship with his employer, but he also saw it as an opportunity to gain leverage over upper-caste villagers and the village elite by professing shared interests with them and building relationships with these and other potential allies in the Sanghatana. At the same time, he continued to pursue caste-specific interests through participation in the CHR. Balu publically performed this shared interest through his participation in the rally, as well as through the invocation of Sanghatana idioms of identity such as the Bali Rajya gav—while fully recognizing that his performances have both "on stage" (Goffman 1959) and "hidden transcript" (Scott 1985, 1990) layers of meaning. In addition, Balu viewed the opportunity to record complaints at the village's Q kendra as a validation of his perception that the Sanghatana could be a potential ally. He saw this as a concrete resource for the protection of his interests, at least in some contexts of village life.

Balu was not the only resident of the Maharwada to describe rationally considered benefits from attendance at the rally. Several other Mahar informants also spoke of the opportunity for bridge-building, as well as the Sanghatana's anticorruption "Q" initiative, in similarly well-considered terms. Moreover, two informants also mentioned that they were interested in learning about the Sanghatana's support for the creation of a separate Marathwada state, independent of Maharashtra—a move that, they felt, would lead to significantly improved representation for Mahars and other Dalits, who are a proportionally larger population in the region.[15]

Photo 8. Shetkaris sowing seed on fertile land in Marathwada.

What we see then is that, for these informants, there are rationally considered interests to be served through at least situational engagement with the Sanghatana and a show of attendance at the rally. Rounding up the weakest sections of the community and transporting them to the site of a rally may be coercive in some respects, but what is lost in that analysis is these participants' capacity to assess in their own terms the significance of getting into the truck and attending the event. It is a negotiated choice that may include, but also goes beyond, conceding to authority or hoping for a free meal. It represents a real opportunity to build political and community linkages. It also represents a potential opportunity to build reciprocal relationships with power that extend outside of the village confines, thus moderating power within the village itself. Just as importantly, it represents a chance for participants to learn more about what the movement may have to offer, and it provides a potential opportunity for them to insert their own voices into the movement, encouraging it to be responsive to specific interests that are relevant to individual participants and their segment of the community.

Photo 9. A shetkari couple on a very small plot in Western Maharashtra harvest their hail-damaged peanut crop by hand.

At the same time, Balu's involvement with the Shetkari Sanghatana, and that of other residents of the Maharwada, is probably not the kind that drives participants to engage in the most risk-laden activism, such as scaling electrical towers or confronting police. Clearly, there are different types of participants even within "participating" villages. Moreover, just as villages may have fluctuating involvement with the movement, individual participants may have fluctuating involvement as well, in step with their changing need to pursue other objectives and their changing perceptions of their interests and opportunities.

Shetkari subjects and shifting participation
The examples of Perugaon and Dorlapur force us to problematize participation in a mass movement, and to think in terms that delve beyond the received constructs of temporality, participatory engagement, and identity solidification. First, they give us some insight into the range of opportunity structures and constraints that can either promote or inhibit participation. As we have seen, participation may be either promoted or discouraged by local wielders of power and influence. These may be, for example, the local dominant caste or class, the local panchayat head or political patrons, the local family of influence, an employer, or a confluence of all of these. At the same time, participants may pursue their interests by situationally aligning themselves with these nodes of power, or through other movements, political parties, or other demand groups. Second, they suggest that overall growth in popular participation in a movement, though cumulative in sum, is not necessarily continuous at the village or individual level. In other words, though the total number of participating people or villages may rise or remain steady over the course of a movement's evolution, we cannot assume that the actual people engaged in the movement—or the intensity of their engagement—are the same from one point in time to another. Individuals and entire villages may come and go, and perhaps come and go again. The third thing we can learn from these examples is that the overarching objectives and subjectivities articulated at the central level of the movement are not necessarily those that constitute the felt purpose of participation, or the actual experience of identity, at the ground level of the movement. Finally, these examples show that participants themselves play a pivotal role in deciding when and how the Sanghatana can help them to advance their interests, and in what contexts individuals are willing to express solidarity with the movement through the

symbols and idioms of movement identity.

At this point, it may be useful to consider how representative these differential interests and varying intensities of engagement are in the movement as a whole. To what extent is this typical?

As mentioned in the previous chapter, at several points in my fieldwork I had the chance to conduct random surveys of attendees at a number of Sanghatana rallies of varying sizes in different parts of the state. Among the structured and unstructured questions that we posed were several aimed at determining the extent and duration of the attendee's involvement in the movement, as well as the specific issues that attendees viewed as most important to them. At every event, responses regarding the motivating issues for their attendance initially showed great consistency. If, for example, the event was publicized as being primarily about prices on a certain crop, respondents generally began by saying that they had come to participate in the event for that very reason. Similarly, if the event had been publicized as being about the regional statehood question, or the anticorruption campaign, respondents would initially speak of these issues as explanation for their attendance. In other words, participants generally defined their participation in terms of the big idiom of the moment, expressing their solidarity with the assembled collectivity.

When pressed to give more information, however, respondents also identified other issues that were often more important or more relevant to them, and about which they hoped to hear the Sanghatana's position. These other issues included a wide range of concerns, including prices for other crops, difficulties facing debt-ridden households, issues related to alcohol abuse and domestic violence, concerns about village water supplies and electrification, or complaints about local schools. Often, respondents indicated that, though they did not expect their concerns to be addressed during the particular event, they hoped that they would have an opportunity to communicate their problem to specific leaders or to others within the movement. These more open-ended explanations from respondents suggest that, even though they were willing to perform a degree of solidarity around the issue that defined the event (by attending and by expressing this as a personal concern), the specific issues that motivated their attendance (and that might lead them to further or deeper participation) may have been quite different.

If the issues of concern varied widely, attendees' responses to questions about length of participation were even more revealing. For example, at the Akola rally in 1996 (cited in chapter 3), the attendees we surveyed identified the length of time that had elapsed since their first attendance or participation in a Shetkari Sanghatana event as ranging from zero months (in other words, this was their first Sanghatana activity) to twenty years. Only a quarter of the attendees surveyed indicated that the elapsed time had been five years or less, and more than half indicated it had been ten years or more since their initial involvement with the movement. This may seem to suggest significant continuity of membership over time. However, when queried further with regard to how *often* they had participated in events over that period of time, most respondents indicated that their involvement had been either periodic—alternating between stretches of deep engagement and relative detachment—or merely occasional.

This points toward the conclusion that a large proportion of the Sanghatana's base of support is made up of "participants" who do not always participate. Moreover, when they do participate, they often do so with objectives and interests in mind that are quite different than those discernable from the "loud voice" representations of movement leaders.

Of course survey techniques, as we have already noted, do have their limitations—but these results definitely correspond with the overall picture of the movement that emerges from my interviews and observations of participation in other contexts of research. They not only correspond with village examples that we have discussed above, but they are strikingly similar to movement leaders' own understandings of their participant base. As a case in point, we can consider the comments of a man I will call Rahul,

174 Cultivating Community

a well-informed Sanghatana leader working in one district of Marathwada. During a conversation with Rahul, I asked him how many people he could call on in his district for organizing efforts and agitations:

> **Rahul:** There are about 700 or so workers who have done *"tapascharya"*—that is, twelve steady years of service in the movement. We just recently recognized and honored these workers at a Sanghatana function. These are the solid, badge-wearing workers. Then there are the on and off workers, who total about 2,000. Then there are the occasional workers, whose participation is strictly issue-based, and who have other political ties. These folks back away from the Sanghatana when it challenges their political groups. These workers total about another 2,000.

> **Author:** So that means, in this district, about 4,700 people across these categories?

> **Rahul:** Yes. And after these categories of "workers," we could speak of "regular sympathizers," who would number about 25,000 to 30,000 across the district. Mind you, this is one of the weaker districts in the region.

> **Author:** How do you reach these estimates?

> **Rahul:** On the basis of attendance at gatherings in different parts of the district. At a recent Mahila[16] meeting, for example, we drew 10,000 to 12,000. If [Sharad] Joshi hadn't been there, it might have been 2,000 to 3,000 less.

Assessments such as these square well with my other interviews and observations with other participants. As I indicated at the beginning of this chapter, my fieldwork with the Sanghatana convinced me that the bulk of participants drift in and out of engagement with the movement and profess allegiance to other arenas of opportunity in other contexts. These shifts may vary tremendously in their temporality. In cases exhibiting the most intense and sustained engagement, shifts toward the Sanghatana may occupy significant passages of time, but may also be followed by effective disengagement for other extended periods. A second pattern is one in which the shifts are more routine, tacking back and forth between contexts of interest and performances of collective identity that are more or less concurrent. Neither of these patterns is necessarily exclusive of the other.

As an example of the former, we can consider the story of a man we will call Ananda, a Sanghatana participant whose shifts occurred over the span of several years. When I met Ananda he had recently rejoined the movement after an extended absence. He explained how he had become involved for the first time:

> I became interested in the Shetkari Sanghatana
> early on. At first, I attended some meetings…but
> always I kept off to the side and did not fully
> participate because I was unsure about this
> movement and the people. I had been involved in
> other movements before, and had learned not to
> trust easily.

Here we see a situation not unlike that of Balu in Dorlapur. Ananda attended a few meetings and was willing to be perceived—to some degree at least—as a participant. Although he had not yet decided to become more deeply engaged with the movement, his presence no doubt conveyed at least a tentative measure of common interest and shared identity with the other people in attendance. Sometime later, however, Ananda made a decision to become more involved and to perform his solidarity with the Sanghatana in a wider array of social contexts. Ananda continued:

> Then in 1982 the shetkaris' situation was so bad
> and I joined in a rail roko [blockade of train

traffic] agitation. I was arrested and was put in jail for three days. By this time I had taken the Sanghatana seriously, and continued to get involved whenever there was a crisis for the shetkaris. Several times, we were beaten by the police for different agitations, but we just called out cheers like "Shetkari Sanghatana zindabaad!" and never ran.

Despite this deep engagement for several years, Ananda then publicly and performatively disengaged from the movement:

Author: And you have been with the Sanghatana ever since?

Ananda: Oh no. There was a period—from 1988 to 1995—when I was involved with other things. There was a dairy scheme launched by some people in [name of village] and my family tried to have some luck with that. We acquired some animals, produced milk…but it was difficult, and there was no possibility to recover even what we were spending. During this time I had no contact with the Shetkari Sanghatana. I felt that those people were against some of the Sanghatana's ideas and that it would be better to keep some distance. But in 1995 I saw that it was a good time to join with the Sanghatana again, and soon I was engaged in organizing villages throughout the taluka [district subdivision].[17]

When I met Ananda in 1998 he was a committed activist with the Sanghatana who had first become engaged with the movement fifteen years prior. In the early years, his degree of commitment entailed confrontations with the police and going to jail. In 1998 his commitment seemed just as strong. Nonetheless, there was a gap of seven years during which he had not participated at all. The reason, as he explains it, is that during that time, participation would have conflicted too significantly with other opportunities for pursuing his interests.

The second pattern of engagement is similar to Ananda's earliest days of attending meetings in which the shifts are more routine. That is, ebbs and flows of engagement that are more frequent and even overlapping with other arenas of identity and participation. This pattern is not only common to fringe participants such as Balu, or Ananda in his earliest, most hesitant days of attending meetings, it can also be easily seen among participants who may be deeply, unhesitatingly, engaged in some aspects of the movement but not in others. For example, in 1997, while visiting a village in Vidarbha, I sat in on a late night meeting with half a dozen local Sanghatana organizers. In this meeting, the primary topic of discussion concerned strategies for expanding the Sanghatana's anticorruption movement in their district. But as the meeting began to wind down, the conversation took a turn. Three of the men in attendance began discussing the electoral prospects of pro-Sanghatana candidates in the upcoming Lok Sabha[18] elections—particularly the prospects of a man from a nearby village who had considered running under the banner of the Sanghatana's unofficial political platform, the Swatantra Bharat Paksha.[19] At this point, one of the participants suddenly excused himself from the meeting.

Several days later, while sitting with many of the same people, the man who had excused himself explained to me why he had left. In addition to being deeply interested in the Sanghatana's anticorruption initiatives, he was also a committed organizer for the local committee of the Congress Party. Although he described his engagement with both the Congress and the Sanghatana's anticorruption campaign as if he were fully committed to each, he also explained that, in such moments where his loyalties conflicted, he either had to refrain from getting involved or to choose one loyalty over the other. When the conversation at the meeting turned to the upcoming elections, in which Sanghatana-supported candidates would be directly competing with those of his own party, the activist recognized that it was time to (figuratively speaking) switch his badges.

Just like the village-level examples of Perugaon and Dorlapur, each of these cases demonstrates different patterns of shifting engagement. The picture of the movement that emerges from this closer look is a dynamic collectivity of complex individual actors rather than a solid body of participants unified in identity and motivation. What we see is akin to what Italian sociologist Alberto Melucci calls a "composite action system" (cf. Melucci 1983, 1989, 1996). It is a confluence of broad ranging actors who, though they do not necessarily agree on the exact nature of their interests or the precise boundaries of their solidarity, coalesce into a pattern of engagement that is more or less stable. On the macro level, this relatively stable pattern of cumulative engagement is what we understand as "the movement." At the individual level, however, we see that each contributing participant may take an active interest in the Sanghatana on the basis of just one issue or a plethora of issues. They may participate at widely different levels of engagement, and at different points in time, alternating with other calculations of personal interests and other opportunities to pursue them. This shifting in and out of professed identity with the movement is very similar to the contextually shifting boundaries of castes and rural classes that we have already discussed in other settings of Maharashtrian village life.

Sanghatanas and sanghataks: Dialogical crafting of agency

If the movement is characterized by a range of differently positioned actors who shift in and out of their experiences and performances of engagement, then one of the major challenges faced by mass movement leaders is to maximize the number of contexts in which wide ranging actual and potential participants correlate their own interests and identity with the demands and ideology of the movement. This also entails the maximization of positive sanctions and rewards that individuals receive for their involvement.

In the previous chapter I have considered some of the ways the overarching objective and ideology of the Sanghatana is crafted by movement leaders to signify inclusiveness and unity for the greatest number of potential participants. In this chapter, I have looked more closely at the shifting nature of participation and the dispersed and evolutionary collection of the movement's declared goals—the whole range of issues, campaigns, and demands undertaken by the Sanghatana at statewide or local levels—that fall within the broad rubric of shetkari unity and the One Point Plan.[20] This evolution of objectives within the movement constitutes a second dimension of strategies through which movement leaders attempt to reach and hold the greatest number of participants in the greatest number of contexts. But how do movement leaders know which goals to undertake in order to maximize engagement? How do they know which issues, campaigns, and demands will create the greatest number of contexts and opportunities for triggering mass participation? In the case of the Sanghatana, this is, to a very large extent, communicated to leaders from the movement's actual and potential participant base.

Shifting participation and participatory agenda setting
Although it would be a mistake to view the Shetkari Sanghatana as an unmediated forum for the expression of the popular will, it would be equally erroneous to look upon it as a strictly top-down dictation of objectives and meaning. The determination of issues to pursue and strategies for pursuing them does not occur in a vacuum. Even if we accept that the movement's central leaders may have a primary interest (as "rich farmers") in the establishment of higher prices and the ultimate free-marketization of the rural economy, we must remember that this objective cannot be reached without adequate mass support. And even if we accept, as I have argued, that the objective of price and marketization may have substantial mass drawing power in much of rural Maharashtra, we must keep in mind that in order to keep the greatest number of participants engaged in the movement in the greatest number of contexts, the movement must facilitate a

dialogue on what *else* actual and potential participants care about and perceive to be immediately important to their livelihood.

Photo 10. This village dry-goods shop doubles as a local Sanghatana "Q kendra" where villagers can register complaints about instances of corruption or harassment. Notably, the name of the shop on the signboard is "Bali Raja Store."

How are issues and potential demands communicated up to leadership from the grassroots? One mechanism of this process that I have already touched on is participation itself. Because the overwhelming share of participation is shifting and inconsistent, leaders must closely observe *which* issues and *which* demands or initiatives draw the greatest response from the greatest number of people. As a rough indication, they may do this is by gauging the number of attendees at an issue-based rally, as described by Rahul. However, we have also seen, in the comments of both leaders and participants, that this might not be the most accurate reflection of an issue's true drawing power. Probably the most accurate way leaders make these assessments is by looking at the more local level responses of participants in the day-to-day process of village and individual organizing. Leaders and local organizers are keenly aware of an issue's significance when, for example, the landholders of Perugaon respond to a joint-stock initiative, or the cotton growers of Dorlapur remain tightly engaged on the issue of cotton prices. They notice the response of a village's Dalit community or other marginal subjects to, for example, the call for regional statehood. They notice when thousands of villages respond to the campaign to elect women panchayat members or to transfer land titles to male shetkaris' wives, and they notice when a worker of the Congress Party (long portrayed by Sanghatana leaders as a tool of the black British) aligns, at least situationally, with the Sanghatana on the initiative to root out local corruption.

But if this were the only mechanism in the dialogue, it would still appear to be more or less a "top-down" affair in which leaders *invent* issues or demands and vet them with the masses. What we need to keep in mind is that these issues and demands are formulated in the *process* of a dialogue, not simply in the initiation of one. In other words, long before these ideas and goals are pitched to actual and potential participants, leaders and local organizers have already listened carefully for insights into the key concerns of agriculturalists in their varied pursuits of livelihood. Let us consider how the objectives and demands articulated by leaders at the central level are informed by the Sanghatana's mass base. The primary way that this occurs is through their extensive contact directly with local participants.

Even in the first years of the movement, Sharad Joshi and other Sanghatana leaders demonstrated an interest to learn about relevant concerns and issues from their potential supporters. Recall, for instance, the former organizer Pramodh's description of how Sharad Joshi communicated with rural Maharashtrians and picked up the "rural wisdom" in the early years of the movement by going from place to place collecting small numbers of people to talk under a big tree or at someone's house. Such intimate meetings, still conducted at the time of my research by a more dispersed group of local organizers, deeply informed Joshi's own understanding and assessment of rural needs. Joshi himself relayed to me an important discovery that he made during one such gathering in a village near his own land. While visiting the village, Joshi learned from the men in attendance that village women had to walk a long distance, several times a day, to collect water from a well. So, after considering several other ways to build bridges with the people in the area, Joshi proposed a plan to construct a water tank close to the center of the village. In public, the villagers were enthusiastic—but secretly, a number of women from the village approached Joshi and begged him not to go through with the plan. As Joshi described it to me "They said 'Please, don't do this; do anything else for the women, but not this.' And they explained that going to the well was the only chance they had to talk amongst themselves, to discuss their problems together."

In this incident, Joshi learned about the men's own assessment of what was needed to help the village—but he also learned that, even in simple matters such as a water tank, interests may differ across the village community. It was a lesson about rural culture and social relationships that he could not have learned from simply dictating to villagers what it was that they needed.

Other informants have described Joshi's capacity to listen and learn from participants in other contexts. One state-level leader who we will call Kishore, speaking of the Sanghatana training manual mentioned in chapter 3[21], described the lectures on which the book was based:

> **Kishore:** Do you know that book's history? It is from a series of lectures, given over three days, at a workshop in Ambejogai, Beed district. There were about 300 people there, none of them probably educated beyond the 7th standard. In those days, Sharad Joshi had only been in agriculture for two years, and so the lectures were sort of more like a conversation. You see, everything he said, he would watch the reaction—he had his ideas, and the education,

but the audience had the real experience, and so he spoke each sentence in consideration of how they responded to the first. He was thinking as he spoke...more like, they were all thinking, and he spoke what they were all thinking. It was like "the group from one mouth."

Author: And did people understand that this is what was happening?

Kishore: In some cases, yes, because there were opportunities for actual discussion, and he would incorporate what they discussed. In other cases, they may have only realized that what Sharad Joshi was saying was very much what they themselves knew from experience. The ideas in that book became famous. Some of the important phrases—like "the cow's mouth belongs to the shetkari, but the udders belong to the state"; "mother can't feed you and father won't let you beg"; "loot them when there's scarcity, and auction them when there's plenty"—these phrases became known by all the shetkaris. They painted them on village houses and walls, using whatever color they could find—red clay, lampblack, calcium carbonate. Joshi may have said these things...but the ideas and experiences were basically their own.

Another informant, a woman we can call Anuja who had been closely involved with the Sanghatana for several years, gave the following description of Joshi's interaction with village women during the founding of the Mahila Agadhi (the Sanghatana's women's front)[22]:

After they established this organization [the Sanghatana], which then included mostly men, Sharad Joshi set out to diversify it to include women. He started having small meetings of

women in the rural areas, and he would be the
only male worker of the Sanghatana present.
Because there were no other men around, the
women would speak openly...they would feel
they were talking to a brother about their
problems. In these women's groups, the women
call him Sharad Bhao [Sharad Brother], as
opposed to Joshi Saheb, as he is usually called by
the men. So he picked up their ideas, transmitted
these to them as his own, and made these
women's concerns an important part of the
movement.

Here again, we see Joshi engaging in a direct dialogue with
participants at the grassroots level of the movement, drawing
their ideas and concerns into the larger process of setting the
movement's agenda. The two-way nature of these encounters
between leaders and participants was by no means limited to the
earliest years of the movement or to interactions directly with
Joshi. On the contrary, opportunities for dialogue have become
institutionalized within the movement and are an expected
characteristic of many of the Sanghatana's most routine types of
gatherings.

While top leaders pay attention to how participants articulate
their interests and, in turn, respond to leaders' own articulations,
the real work of participatory agenda setting occurs on the most
local level of routine organizing and participant interactions.
Because of the loud central voice of the top leaders, this dynamic
is sometimes difficult to see. Consider, for instance, a typical
Sanghatana rally. The rally may open with local leaders who all
address the primary issue for which the rally has been
convened—say, the price of cotton, or the campaign for women's
land inheritance rights, or the question of Marathwada statehood.
After each of these speakers has addressed the issue from local
and regional perspectives, the podium is given to the most
prominent leader in attendance—perhaps Sharad Joshi himself—
who then discusses the particular strategies and demands that the

movement intends to pursue in relation to the issue, as well as the significance of the issue in the context of the overall philosophy and agenda of the movement. To the observer, this may appear to be a simple dictation of what the movement is and what its chosen issues and demands may be at any point in time. What is lost in that perception is the way that these ideas have percolated up to leadership long before the rally has been called.

Photo 11. Landholders register as shareholders in a joint stock company for leveraged sales of land to government or private developers. Here, Sharad Joshi personally hands over the completed shareholder certificates to a participant during a registration session at Shetkari Sanghatana headquarters.

To understand the pronouncements at the rally, it is important to consider that each of the leaders appearing at the podium have formulated their own assessment of the movement's potential goals based on interactions with their own local or regional constituencies. These leaders derive their understandings of important issues and demands from conversations with local level organizers and activists who, in turn, derived their own understandings by working closely with current and potential participants. By the time issues are raised to the level of Sharad Joshi or other senior leaders, they have already been well-considered in village and regional meetings and assessed for their potential to recruit and hold participants. Moreover, Joshi and other principal leaders reinforce this process of upward communication within their rally speeches by telling attendees to talk with their local organizers. As Joshi told the audience at the end of one large Mahila Agadhi gathering, commenting on an upcoming massive rally scheduled for the following month: "This is *your* gathering, so let us know what you want to talk about there." In other contexts, I have heard Joshi tell delegates that he and other leaders should be regarded as peers or fellow participants rather than as unapproachable authorities. Speaking at a small rally that I attended near Perugaon, on the subject of creating an organized response to the MIDC land acquisition plan, Joshi exhorted an audience of approximately five thousand attendees to address him in the familiar manner of a peer rather than with honorific terms: "Today you may consider me your leader and you may be ready to do everything for Sharad Joshi. You may be willing to die, to go to prison, to face a lathi charge...so let me make certain things clear. I am Sharad Joshi. Not 'Sharadrao.' Not 'Sharadraoji.' And definitely not 'Sharadraoji Joshiji!' I am one of you."[23] He then went on to say that for the movement to be successful everyone would need to consider themselves equally important contributors: "If we all come together, we will solve this problem. We [the Sanghatana leadership] will come whenever you ask, and we will do whatever you tell us needs to be done."

Most of the back and forth communication that underlies participatory agenda setting is informal, occurring in mundane village interactions and routine conversations every day. Sometimes, however, the feedback loop is formalized in deliberate meetings aimed specifically at building bridges between ground-level participants and the movement leadership. As the most formalized example of the latter, we can consider the Sanghatana's annual *karyakarini* gathering—a sort of constituent assembly where central leaders, regional leaders, a range of district- and taluka-level organizers, and ordinary participants meet to engage in a dialogue. These gatherings are forums for discussing the state of the movement, its achievements in the past year, and its plans for the year ahead.

At a two-day karyakarini gathering that I attended in 1996, the program was structured to encourage the greatest possible input on what the movement should undertake in the upcoming year. Although the second day of the meeting was reserved specifically for recognized leaders and organizers, the first day of the meeting was set aside for open discussion, with participation open to anybody who had an interest in the movement. For several weeks prior, news about this opportunity to attend an open forum with Sanghatana leaders was broadcast throughout the state by village organizers and published on the back page of the *Shetkari Sanghatak* newsletter.

During this first day, all of the attendees had an opportunity to speak before the group to report on their key concerns and to make suggestions about issues that the movement should address. At least a hundred people got up to speak at different points during the day—some of them men, some of them women; both Hindus and Muslims; people who were clearly educated as well as people wearing the worn out saris and dhotis of the poorest villagers. Each of them, speaking for about ten minutes per person, seemed to have carefully considered what it was that he or she wanted to relay to the group; many of them dug into their pockets or saris, pulling out carefully folded notes to guide their remarks. After the formal hearing was over, groups split up to take tea and stake out sleeping spots in various corners on the floors of the meeting hall; conversations continued well into the night.

The second day of the program was limited to committed activists and officeholders in the movement—about five hundred people—and entailed a well moderated group discussion of the presentations and issues that had come up during the previous day. The program also presented opportunities for workers from all the different active areas of the movement to report on the ground level activities in which their local groups were engaged and the various issues that shetkaris in their own villages and talukas faced in the year ahead. As the day before, individual men and women stood up before the group to make their short, but well considered, addresses. Typically, these addresses began with an opening introduction comprising some variation on the following:

> My name is _____. I have been working for _____ years with the Shetkari Sanghatana, in the struggle for a fair return on the shetkaris' sweat and labor, from _____ village, in _____ district, in the land of Bharat, kingdom of Bali.

After this, speakers might mention some of their key achievements in the movement, or important activities in which they had played a role—essentially, presenting his or her credentials—before getting to the real substance of the important issues facing their constituents. The issues mentioned were very often distinctive to the speaker's local area and significantly outside the strict letter of the One Point Plan. One man, for example, after declaring his commitment to remunerative prices and the restoration of Bali's rule, proceeded to describe a village in Chandrapur district that was at risk of being taken over by the government to make space for the expansion of an airport. The Sanghatana, he felt, should consider ways to help. By the end of the day, some of the local issues raised had been embraced by delegates as things that the Sanghatana should address at the state level, through statewide initiatives, demands, and agitations. Others were endorsed as things that the local chapter could pursue on its own, with recommendations shared from other leaders about how the local workers might best approach the problem and organize a response.

Forums such as this suggest that, by the time senior leaders address crowds at massive rallies and help to organize activities in any particular part of the state, their articulated goals and actions are deeply informed by local assessment of issues that are relevant to participants at the ground level. But this is only the most formal example. Most of these feedback opportunities occur in ordinary village settings and mundane, chance conversations. Charged with the responsibility of organizing as many people into participation as possible, village and regional workers are constantly listening for issues and concerns that can deepen the engagement of current participants and draw new sections of their communities into the movement.

Photo 12. Villagers gather late at night to meet with an arriving pro-Sanghatana candidate for the Lok Sabha elections. Lit by the headlights of the vehicles in his entourage, the candidate made a speech in the street and discussed local issues with attendees before heading on to the next village.

These are mechanisms through which laborers, for example, were able to communicate their concerns to Sanghatana workers. Undoubtedly, this upward feedback strongly influenced Joshi's calls, beginning around 1984, for labor-employing landowners to pay a fair daily compensation that exceeded the government's own established minimum wage for agricultural labor.[24]

Informants have explained to me that this is also how women communicated to Sanghatana leaders, through the channels of the Mahila Agadhi, that female rights of land inheritance and opportunities for women to market surplus garden produce were key issues that affected a wide spectrum of rural Maharashtrian women. Far from being one-direction dictates by the most empowered central leaders, most of the Sanghatana's practical ground level initiatives—whether wages for labor, land inheritance, marketing co-ops, corruption eradication, debt relief, organizing against government land takeovers, or setting price demands on specific crops and other produce such as milk—can ultimately be traced to some manner of dialogical engagement between leaders and the movement's actual or potential participant base.

Dialogues of interest and movement ambiguation
These dialogues of interest and meaning that occur throughout the movement may not always be easy to discern. Part of the reason for this is that much of it begins in quiet and uneventful conversations, in seemingly insignificant village settings. But there is another very important reason: when these dialogues *are* public and observable, they are very often conducted within the idiom of the movement that we associate with the "loud voice" representation of its leaders.

This shared idiom may give the impression of consensus. However, when we look more closely, we see that it can also function as a vehicle for exchanging views on just *what* an actual consensus would need to entail, from each speaker's perspective, in order to be most acceptable. When the delegate from Chandrapur addressed the audience at the karyakarini gathering in the Sanghatana idiom—invoking the kingdom of Bali and the struggle for shetkaris' return on their sweat (in other words, remunerative price)—his invocation served to express solidarity with the collectivity and to validate these established signifiers of the movement. At the same time, however, by raising the issue of the airport expansion and the village takeover, he sought to insert a new issue, a new set of interests, and a new dimension of meaning into that constellation of signs. Likewise, when Sunita likened the fate of Bali to the threat of industrial expansion facing Perugaon, she asserted that Perugaon's interest in forming a joint stock company was consistent with the meaning and objectives of the Sanghatana. And when Balu invoked Bali Raj in conversation with Dorlapur's large landholders—all the while referring, in his own mind, to justice for the residents of the Maharwada—he is subtly signifying that the ideal of Bali's realm embraces the demand for Dalit equality and the interests of labor. Each of these agents, speaking in the idiom of Sanghatana solidarity, are engaged in a struggle to define the range of meaning that adheres to each of its signs.

Thus, while movement leaders are engaged, at one level, in constructing an idiom of the movement that signifies shetkari unity around the One Point Plan, participants and leaders are both engaged in a dialogue on the full meaning of that idiom and the broader mass interests that it represents. The capacities of leaders to hegemonically craft the movement's agenda and collective identity, therefore, is not only limited by the selective participation of variously positioned rational participants, it is also limited by leaders' need to attract and retain participants by actually promoting these dialogues. The movement must differentiate itself from other opportunities in the agentive landscape, but it must also evolve to address emerging challenges to subjects' livelihoods and to accommodate new terrains of interest and identity.

One way to think of this is in terms of a distinction that information theorist John Leslie King and historian Robert L. Frost (2002) have drawn between strategies of *disambiguation* and strategies of *ambiguation*. Disambiguation, the strategy most commonly emphasized in studies of organizational goals and ideology, is what they define very simply as "making clear what is meant and intended" (4). King and Frost define *ambiguation*, on the other hand, as "the effort to keep meaning and intent vague," and they argue that this strategy, while far less explored by theorists, is at least equally important to understanding the success of dispersed organizations and ideologies (Ibid.). For example, in their discussion of the Roman Catholic Church during its historic spread through different ranks of society and across great geographic expanses of cultural difference, King and Frost suggest that:

> ...it is tempting to ascribe the success of the Church to the establishment of highly rigorous and strictly enforced hierarchy of authority and belief, reaching down from the pope to the most remote parish priest. In this model, the doctrines of the Church were articulated precisely and disseminated down through the hierarchy. A closer look at the history of the Church during its great evangelical expansion reveals that institutional survival was very much a result of a carefully constructed and maintained doctrinal ambiguity (11).

The significance of this doctrinal ambiguity, according to their analysis, is that it permitted a wide range of people across a broad social and cultural spectrum to interpret the meaning and objectives of the Church in their own way, and to communicate back to local Church authorities what it was that they expected from their engagement.

In the Shetkari Sanghatana we see active attempts at both disambiguation and ambiguation. Disambiguation is evident, for example, in leaders' use of simple, recurring terms and phrases to articulate what the movement stands for and how it is differentiated or continuous with other struggles and ideologies. But the Sanghatana also provides wide allowance for ambiguity by deliberately avoiding tight definitions of these terms. This creates space for dialogue and enables the movement to be receptive to other objectives, other constituents, and other terrains of identity. It enables the movement's opportunity structure to actually be remade in the practice of participation.

Ambiguation and Sanghatana organization
The classical understanding of organizational evolution tells us that, as organizations expand and their social presence becomes fixed, they become increasingly formalized and bureaucratic. The Shetkari Sanghatana represents an opposite possibility. As the Sanghatana became an increasingly popular and influential movement, it also became in many ways an increasingly *less* formal and bureaucratic organization. For example, despite the Sanghatana's impressive capacity to communicate the "loud voice" version of its goals and ideology through popular media channels, its organs of communication and governance have over time become more dispersed and less formal.

When Dinkar Deshmukh (1985), then an MPhil student at the University of Pune, conducted research on the Sanghatana in the early and mid 1980s, he found the framework of a swelling bureaucracy in progress. In addition to a statewide president, three vice presidents (one each for Western Maharashtra, Marathwada, and Vidarbha), and "district correspondence workers" (one for each district in the three regions), he also identified a range of formalized departments and special committees tasked with the functions of movement governance, coordination, research, and propaganda. Each of these had tightly delineated responsibilities. The Propaganda Department, for example—though it only retained five staff members at the time—was already organized into twenty-three separate divisions ranging from "training" to "cultural activities" to "youth." The Agitation Department had three separate divisions, staffed for the coordination of different types of agitations, and the Agricultural Produce Department had six divisions for conducting research and analysis, ranging from "foodgrains" to "sugarcane" to "onion" (148–149).

Photo 13. After staging a small-scale agitation on crop prices, sanghataks from several nearby villages gather to discuss the success of the day's activity and to strategize future activities in the area. In the process, they discuss other problems and concerns that affect their village communities.

By the time I conducted my own fieldwork in the mid and late 1990s, the Sanghatana had become a vastly less formal organization. All of the departments identified by Deshmukh had vanished—and, despite the fact that the movement had launched a wide range of new programs and initiatives in the intervening years, no new official departments had been established at the central level. The movement's initiatives were coordinated and managed by volunteers across the state and not overseen in any formal manner.

Of course, central leaders were deeply engaged in the movement's activities, but the authority of their leadership had proven equally resistant to formalization. Although the Sanghatana continued to have a state president, regional vice presidents, and district workers (as well as parallel positions for each of these in the Mahila Agadhi, added in 1986), no new formalities or legitimizing structures of their leadership had been created. The movement had no formal constitution or constitutional rules for decision-making, and operated as before on the basis of consultation and consensus.[25] Even Sharad Joshi, though the universally acknowledged charismatic leader of the movement, continued at the time of my fieldwork to hold no formal office and claimed no official grounds for his authority.

Perhaps most significantly, the Sanghatana had given up all efforts to register its members. At the time of my field research, the Sanghatana had no formal rules or rituals of membership. Anyone who participated in any way, if they choose to do so, was free to describe himself or herself as a member—and was not required to give up any other formal or informal memberships in order for their claim to be accepted as valid.

Thus, in both form and practice, the Sanghatana's organization was far from the rigid and formal entity that we might expect of a movement that has been attracting participants, crafting agendas, developing a distinctive idiom, and challenging government for nearly two decades. Swedish sociologist Staffan Lindberg (1994), comparing the Shetkari Sanghatana with the Bharatiya Kisan Union (in Uttar Pradesh state), describes the Sanghatana in the following terms: "At times when there is no agitation, it is as if the organization does not exist [...] Thus the organizational form is anarchic or 'post modern' in the sense that, much like the new social movements in the West, it builds structures around actions rather than routine organization" (112). In this sense, the formal organization of the Sanghatana appears just as ambiguous and context-driven as its actual goals, practices, and meaning.

None of this is intended to disregard the actual power that resides with the movement's central leadership to influence the course of the movement. Neither do I intend to discount the role that charismatic leaders can play in leading a relatively informal body of decision-makers toward a real (or merely apparent) consensus. But the movement's minimal degree of formal organization indicates that, by design at least, it is not demanding of hierarchy or resistant to change. It has not formally codified its structures and practices for bottom-up dialogue or for top-down enforcement of the dictates of leaders. Organizationally, it is amenable to its own continuous reinvention.

Conclusions

In the previous chapter, we explored the participant experience of the Shetkari Sanghatana through a macro-level lens, focusing on its economic agenda for change and its popular cultural idiom as these are most typically presented by the loud voice of the movement. Here in chapter 4, I have dug deeper into the first of these, the agenda for change.

In the first part of the chapter, we took a beneath-the-surface look at issues and demands addressed by the Shetkari Sanghatana. I peeled back the outer layers of the movement's agenda to show that, at other levels of action, the Sanghatana undertakes initiatives representing a wide range of issues that are only tangentially, if at all, related to remunerative prices. Where the singular demand of the One Point Plan for pricing gives us an image of the movement that is monolithic and unified, these lower layer issues and initiatives help us recognize the movement as a more complex conglomeration of interests. Next, we took a closer look at participants and participation. Through the examples of Perugaon village and Dorlapur village, we have seen that individuals and communities may shift back and forth toward greater or lesser engagement with the movement based on their perceived interests, their assessment of how well the movement can help them further those interests, and their alternative opportunities for action. These examples help us to understand that participation in the movement is characterized by flux rather than continuity. Based on what we see below the surface, the movement appears less like a solid, unified mass and more like a "composite action system" (cf. Melucci 1983, 1989, 1996).

As perhaps anywhere else, social actors in Maharashtra operate in numerous different arenas of collective solidarity. Faced with a multiplicity of interests, affiliations, and opportunities, the Shetkari Sanghatana seeks to attract or hold adherents through the adoption of new issues and the creation of new opportunities that appeal to its actual and potential participant base. Within the Shetkari Sanghatana, participants have substantial capacity to influence the range of issues and opportunities that the movement represents. This, as we have seen, is largely accomplished through bottom-up mechanisms that enable ordinary participants to help craft the movement's agenda. One mechanism of this process is the fact of participation itself. Because participation is shifting and inconsistent, leaders are able to take note of the issues, demands, and initiatives that draw the greatest response from the greatest number of people. Another mechanism is routine direct contact between leaders and participants, which exposes leaders to local concerns that can be communicated back into the organization. A third mechanism is the structure of the Sanghatana's local and state meetings, which include scheduled open mic time for individuals and community representatives to suggest issues that they would like to see the movement undertake.

In the final part of the chapter, I applied the notion of ambiguity to the Sanghatana. Although the macro-level view of the movement gives the impression of a quite straightforward economic agenda and demands, the field view shows these to be relatively ambiguous, open to interpretation and contestation by participants. This is not a weakness, it is an integral characteristic of the movement that affords dialogue and enables the movement to create alignments and opportunities for a broader range of interests and constituencies. In some respects, this ambiguity is reflected in the structure of the organization itself, which is surprisingly decentralized and informal.

In the next chapter, we will continue to dig beneath the surface, focusing on ambiguity and dialogue in the construction of the movement's popular idiom and the contours of Shetkari Sanghatana identity.

Notes

[1] As of 1998 the SBP continued to be an unregistered party, and hence its candidates must actually run under the election symbol of another party—typically the Janata Dal party. Part of the reason that the SBP has not formalized its party status is that its officers refuse to swear to defend the principles of Indian socialism, as is a requirement of the party registration process. Support of the SBP is not mandatory for Sanghatana participants, and many participants—even relatively committed activists—continue to remain active in other formal, registered political parties, including Congress and the Shiv Sena.

[2] This campaign is predominantly, though not exclusively, directed toward shetkari women growing food on small plots for household consumption. *Sita* is the wife of Rama, from the epic Mahabharata. Sita is popularly conceptualized as the ideal of wifehood, and also the ideal of purity. Together Rama and Sita popularly represent the ideal couple. *Sheti* is Marathi for "agriculture."

[3] From *Lakshmi* (the goddess of wealth) and *mukti* (liberation). The name of the campaign implies that women are the real force of wealth generation in agriculture and yet do not enjoy any of the freedom that wealth should be able to provide for them.

[4] On a more demand-oriented front, the Sanghatana has also been, at least since 1993, deeply involved in legislative efforts to secure a 33 percent guaranteed reservation of seats for women in state assemblies and the national Parliament.

[5] Given the dispersed nature of Sanghatana initiatives, it also follows that not all collective protest activities are pointedly focused on specific demands. One of my favorite examples, overlooked by the press despite its novelty, was the "dancing like Michael Jackson" agitation staged in Bombay in 1997. At that time, a minor scandal erupted when Balasaheb Thakeray, the de facto leader of the Shiv Sena party in Bombay, personally facilitated an agreement with Michael Jackson to perform a concert in the city. At one point, a reporter asked Murli Manohar Joshi, the Shiv Sena party Chief Minister of the Maharashtra state government why he and Thakeray could put so much energy into brokering the Jackson concert and so little energy into remedying the problem of rural poverty. To that, the Chief Minister quipped: "When the farmers can dance as well as Michael Jackson, we will listen to them too." Within days, hundreds of Sanghatana activists converged on Bombay to

humiliate Thakeray by performing a group moonwalk and other choreographed dances in front of his home.

[6] Both village names are pseudonyms. In the following discussions, I have altered certain details in order to disguise the true identities of the villages and informants.

[7] To put this in perspective, the government-fixed land ceiling for unirrigated land in Maharashtra is 54 acres. For irrigated land, the ceiling is 18 acres. In some cases, as is common in many villages, nuclear families may have effective control over a larger quantity of land than the amount they officially "own." This is because the title to additional plots may be registered to an owner's siblings who may not actually reside in the village.

[8] Following Putnam (2000), I use the term "social capital" here to refer to the usable value that individuals derive from social networks. Putnam describes this value as the specific benefits that individuals gain from social networks in the form of trust, reciprocity, information and cooperation.

[9] In a previous situation, cited in the *Indian Express* (November 14, 1996, p. 1), and attributed to comments by Sharad Joshi, agriculturalists near Pune sold their land to the MIDC for 13,000 rupees per acre. The government then sold the same land to industrial developers at 1,300,000 rupees per acre—or 100 times the price paid to the landowners.

[10] In the historical social order of the region, *balute* in the form of grain, garden produce, a small amount of money, or small patch of land was normally given to each of the 12 *balutedar* communities who rendered special services to the village. These service communities ranged from village priests, to various types of craftspeople, barbers, and others whose services were unique and defined by their jati. Mahars, usually residing on the outer edge of a village, were regarded as village watchmen, and generally received *balute* for this service. They were also typically expected to deal with "unclean" chores, such as the removal of dead cattle.

[11] A common flatbread. This phrase "providing the roti" is a common one, meaning that the provider is the source of the receiver's livelihood.

[12] Diwali, also known as *Deepawali* or "the festival of lights," is a five-day festival primarily oriented toward the Goddess Lakshmi, who represents female divinity in her aspect as Wealth. Each day of the festival has different implications and practices that vary in their significance by region and community. In general terms, the final day is reserved for

visiting friends, giving and receiving sweets. For laborers and servants, this final day is marked with expectation of some sort of annual "bonus" from employers, which could be a meal, a gift of sweets, or some small amount of cash. For this reason, the festival can be seen as one that reinforces and validates the hierarchical relations of patron-client ties.

[13] The CHR, launched in 1990, began as a collectivity of 15 to 20 small Dalit rights groups that agreed to engage in a coordinated struggle using the name "Campaign for Human Rights." These organizations, working primarily in the Maharashtrian districts of Beed, Parbhani, Nanded, Osmanabad, and Latur, agreed that the initial focal issue of the campaign would be caste-based oppression and "customary bondage" (a form of labor bondage not based on debts but on socially enforced, caste-based labor, such as the customary Dalit obligation to dispose of animal carcasses and human excrement). The initial target communities for mobilization were the two predominant Dalit castes of the area—Mang and Mahar—which historically have been competitive with each other both socially and politically. In the second year, the CHR expanded the struggle to include the poor/landless of any caste or community, on an economic (not caste) basis. And in the third year, the campaign was broadened further to include such issues as bureaucratic and police corruption. These first three years demonstrated tremendous promise for the movement, and the campaign grew rapidly across Marathwada. By the third year, however, unity within the movement began to collapse. By 1996, the CHR coalition had fractured into two dominant and mutually uncooperative, factions—one retaining the name Campaign for Human Rights and the other taking the name Social Justice Movement.

[14] Cotton grower. From *kapus* (cotton).

[15] It is worth noting here that this is an issue of longstanding concern to the Dalit communities of Marathwada. Dr. B. R. Ambedkar (1891–1956), a key figure of the Independence movement and advocate of Dalit rights, was deeply opposed to the establishment of Maharashtra as a unified linguistic state. Among Ambedkar's concerns was that Dalits in Marathwada and Vidarbha, where their numbers are significantly higher than in other parts of the state, would be underrepresented in the larger state dominated by Maratha interests (see Ambedkar 1979).

[16] Referring to the Sanghatana's Mahila Agadhi, or Women's Front. Attendees at Mahila rallies are predominantly women, but some men (in my experience, generally on the order of 15 to 20 percent) attend as well.

[17] A *taluka* is a district subdivision. Districts in Maharashtra comprise anywhere from 5 to 14 talukas (with the exception of Greater Mumbai district, which has only 3).

[18] The Lok Sabha—or the House of the People—is the lower but more powerful chamber of the Indian Parliament. Representatives are directly elected from constituencies throughout the states.

[19] In the 1998 Lok Sabha election, the SBP, whose candidates ran under the symbol of the Janata Dal party, initially declared more than 20 different candidates for the election. In the end, the SBP ran only 5 candidates from the districts of Nanded, Jalna, Buldhana, Wardha, and Latur.

[20] It is important to maintain a distinction here between the goals of individual participants or interest groups within the Sanghatana, and those that are declared by movement representatives. Etzioni (1964) provides a useful terminology for thinking about the evolution of declared organizational goals, distinguishing between *goal multiplication* and *goal succession. Goal multiplication* refers to the expansion of objectives in order to enroll new participants or to address new social and economic conditions. *Goal succession* refers to the identification of new goals in order to replace previous goals that have been either realized or denied. An example of goal succession would be after a price demand on a particular crop has been realized, thus providing incentive for movement leaders to articulate a new goal of equivalent interest in order to retain participants who may have joined the movement in pursuit of that one objective. Both *goal multiplication* and *goal succession* apply to the Sanghatana case.

[21] Shetkari Sanghatana Vicar ani Karyapadhdati (Sharad Joshi 1982).

[22] The Mahila Agadhi, founded in 1986, is not a separate movement but a parallel structure within the Sanghatana. It has its own central leadership and cadre of workers and organizers, corresponding to the predominantly male leaders, workers and organizers of the Sanghatana in general. Although the Mahila Aghadi may undertake initiatives and stage agitations or rallies that are specifically oriented toward women participants and women's issues, it is an integral component of the movement as a whole. Any reference to the "Shetkari Sanghatana," or any activity undertaken by the Sanghatana, always includes the women's front.

[23] Rao and ji are honorific suffixes. The doubling, and then tripling, of these suffixes onto a person's name is clearly excessive and comical—

emphasizing Joshi's point that he doesn't care for honorifics (and getting a good laugh from the audience).

[24] This also prompted Joshi to give greater theoretical attention to the idea that higher prices and a liberalized economy would ultimately lead to more agricultural investment and greater demand for labor.

[25] At one point, the Sanghatana did draft a constitution, but it has never been adopted within the organization. This constitution was prepared only to meet a requirement of government, which obliges formal organizations to be legally registered and their constitutional objectives declared.

5—Bali will rise: Dialogics of interest and identity, continued

> *We have always thought of this wada as the real place of Bali, and of ourselves as the real Bali Rajas. Who is it that toils?...So, when we are talking with other people we may call the whole village a Bali Rajya gav...but when we do so we are really thinking of ourselves.*

—Balu, Dorlapur village, Maharashtra

Vijaya and her husband Ashok, Deshasta brahmins from Ahmednagar district, demonstrated a household ritual for me. This ritual, they explain, is practiced by brahmin families from their ancestral village during the annual *Dasara* holiday. It is a game-like activity. Seated on the stone floor of their kitchen, Vijaya reached into a large storage vessel of uncooked rice and measured out two handfuls onto two different stainless steel plates. She then carefully fashioned each pile of rice into the shape of a man, whom she declared to be Bali. At this point, after asking Ashok to turn his eyes away from the plates, Vijaya hid a small piece of gold jewelry inside one of the rice figures. All the while, Ashok narrated to me his own interpretation of the significance: "You see the rice represents agriculture, and the gold is wealth. Bali represents not just agriculture, but also the agriculturalists and the wealth that they produce." When she had hidden the piece of gold, Vijaya asked her husband to guess which of the two images contained the symbol of wealth. Taking a sharp kitchen knife in his hand, Ashok probed into the belly of one of the images. He had guessed correctly, and he extracted the gold.

In previous chapters I have considered a number of cultural resources that have been drawn into the Shetkari Sanghatana's constellation of signs. I have considered their role in crafting and reflecting a Sanghatana ideology and a Sanghatana community, as well as their significance to various other contours of identity in the state. Of all the signs of community and movement identity that have been deployed within the Sanghatana, however, those associated with the myth of Bali have been by far the most important. They are the most widely invoked, the most constitutive of an idiom of the collectivity, and the most closely metanymic not only of the movement and its agenda(s), but also of its community, its participants, and its leader. Up to now, however, I have not attempted to thickly describe Bali's import in rural Maharashtra.

In this chapter, I will focus on the mythic king Bali as a complex of diverse and contested meanings across Maharashtra. Approaching Bali as a "metasign"—that is, as a sign capable of reflecting and signifying the movement as a whole—I will examine some of the ways that subjects within the movement use the Bali complex to signify their interests and identities in contest with the interests and identities of the state's urban and rural elite.

As I shall argue, one of the chief reasons for the deep success of the Bali myth in signifying both community *and* difference within the movement is its ambiguity and its consequent capacity for multivocality. In this sense, the struggle of differently situated subjects to define what it is that Bali *also means* is very similar to the competition over everything that is *also* signified under the rubric of the One Point Plan. The result of this ambiguity and multivocality is that the Bali myth is able to function as a core element of a movement idiom that is widely meaningful and embraceable *across* multiple terrains of identity or interest, while it also constitutes a vehicle for dialogue on variations of identity and interest within the movement community.

Bali as a Maharashtrian sign of subalternity

Like all of the other powerful and durable cultural resources in the Sanghatana constellation, the myth of Bali Raj is deeply Maharashtrian and an important piece of the language and culture of the state.[1] Although narratives and practices associated with Bali trace to representations of the demon king in the same ancient texts that inform much of contemporary folklore throughout India, Maharashtra is one of only two states in the country where Bali is a prominent and recurring cultural symbol.[2] In Kerala, the other state in which Bali is a widely celebrated mythological figure, the demon king assumes centrality primarily in the context of the grand *Onam* festival, celebrated annually in rural and urban areas alike.[3] In Maharashtra, Bali lives a much quieter cultural existence. There are no major festivals dedicated to his memory, and urban Maharashtrian Hindus—though generally familiar with Bali's role in one or more mythic narratives—rarely, and at best obliquely, acknowledge Bali within ordinary social intercourse or in the cycle of their annual ritual calendar.

In *rural* Maharashtra, however, Bali is a figure frequently recalled and invoked in numerous community and household rituals, as well as in the normal fabric of day-to-day life. For example, one familiar Marathi phrase that can be heard throughout rural Maharashtra, particularly in the context of specific village rituals, is "Let all evil go away, and the kingdom of Bali be restored" (*ida pida talnar, balice rajya yenar*). Even more significantly, Maharashtrian agricultural and laboring subjects commonly express personal identification with the demon king— so much so that individual agriculturalists regularly greet each other as "Bali," and village folk songs often use the name "Bali" as a reference to village cultivators or toilers. This association between agriculture/agriculturalists and the demon king is widely recognized by non-agricultural communities as well. Historical accounts of Maharashtrian life, extending back to the early twentieth century, note that landlords, moneylenders, government administrators, traders, and rural merchants often addressed village cultivators and laborers as "Bali," and this practice is still encountered today.

On the surface, Bali appears to be a figure with whom all agrarian Maharashtrians might easily identify. But scratch that surface and we see the boundaries of Bali-identity are as complex, contested, and contextually shiftable as the terrains of other Maharashtrian identity markers of the agrarian masses. While many agriculturalists address each other as "Bali" with pride and evident self-identity with the deposed king, the most elite members of rural communities have often deployed the name of Bali as a one of many subtle tools for exercising control over the weakest sections of the village. Trimbak Narayan Atre, a colonial revenue officer in the early twentieth century who documented village social relations with an adept ethnographic eye, describes how rural landlords and brahmins often addressed poor villagers and members of the hereditary service castes by the name "Bali" as a form of strategic flattery. This, Atre explains, was intended to appeal to their Bali-like humility and generosity, in order to extract from them greater degrees of service and subservience.[4] Likewise, these same elite groups, in much of rural Maharashtra, engage in a wide range of ritualized enactments of their social control over village communities—including rituals that specifically dramatize the conquest of Bali, such as the ritual demonstrated for me by Vijaya and Ashok at the beginning of this chapter.

Bali, then, carries an extremely wide range of meanings in rural Maharashtra. Some of them are proud and positive, others servile or derogatory. Many of these complexities of social meaning are reflected in the range of words and phrases associated with Bali that have become concretized in regional vocabulary. For example, the Marathi words *balivard* and *balivahan*, etymologically derived from Bali, refer in usage to a bull or a bullock (a castrated bull). The first means "Bali-helper," and the second translates literally as "Bali's vehicle." These words may carry nuances of toil, strength, castration, or all of these, but are clearly tied to the land, and to the agriculturalists who identify with Bali and use bullocks for plowing their fields. A modern derivative of these words, but one with substantially different rural class implications as well as much greater connotations of strength, is the relative neologism *balivardyantra,* which is used for tractors or bulldozers. The term means, literally, "bull machine"—or, even more literally, "Bali-helper machine"—and many smallholding agriculturalists who identify with Bali see this term as deeply paradoxical. As one informant described it: "The real Bali Rajas are the small shetkaris not the big farmers...so we look at the tractor [balivardyantra] as the *destroyer* of Bali, not as Bali's helper."

While some of the many Marathi words that informants sometimes associate with Bali carry meanings of strength—such as *balishta* (potent) and *bali* (mighty)—common Marathi speech includes an almost equal circulation of terms that connote submissiveness. For example, when the proper name (*Bali*) is used as a non-proper noun, it means oblation, offering, or sacrifice. This is often applied to the actual material given in a religious offering to a deity, such as flowers or items of food—in much the same way the demon king offered his own head as a sacrifice to Vishnu—but also refers to donations made to a brahmin. The verb construction *balidan* (to "give" *bali*) refers to an act of generosity or martyrdom. These words may not all have true etymological connections to Bali the demon king, but what is important is that the connections are felt and are meaningful for many people when they use these words.

Many of these terms are not unique to the Marathi language, but their use in Marathi embodies a distinctive thickness of meaning extending from the prominence of Bali in rural Maharashtrian culture. As their variant connotations suggest, these words are reflective of social contests over the essential meaning of Bali whenever he is invoked in speech and practice. With regard to conventional caste hierarchies, identifying oneself or one's own social group with Bali is the inverse of identifying with a higher tier caste. To express identity with Bali is to express identity with the downtrodden—associating one's own social and economic position with that of the least empowered and least Sanskritized sections of the agrarian masses.[5]

This distinction is critical for our understanding of Bali as a movement-wide idiom, and sheds additional light on the specific geographic spread of the movement within the state. The Shetkari Sanghatana, it may be recalled, has seen its greatest success in the state's eastern regions of Marathwada and Vidarbha. These are the regions that have comparatively high populations of Dalit communities and where the kunbi agrarian castes have weaker cultural and political ties with the Maratha elite than do their kunbi counterparts in Western Maharashtra.

Considering the wealth of common associations that tie Bali to the soil, virtue, and strength—as well as to servitude and downtroddeness—it is not surprising to find that the story and figure of Bali have more than once been adopted as ideological signs for rural agrarian social organizing. Very notably, as discussed in chapter 2, the social reformer Jyotirao Phule promoted Bali as a central symbol in the rural Satyashodhak movement during the late nineteenth century. The Satyashodhak movement, which was most successful in the Western Maharashtrian areas historically dominated by the brahmin-Maratha political and ideological compromise initiated by Shivaji, appealed specifically to the most disenfranchised shudra and "ati-shudra" (Dalit) sections of rural society. Bali has also been invoked in Western Maharashtra within a number of other smaller Maharashtrian agrarian movements and political organizations targeted at rural laborers. One example of this is the agricultural laborers' movement called the Shetmazur Shetkari Sanghatana (no relation to the Shetkari Sanghatana), confined mainly to four heavily irrigated sugar-producing districts of the Desh subregion.[6]

Bali's most prominent and massive use in twentieth-century Maharashtra, however, has been by the Shetkari Sanghatana—a movement that seeks the participation of all rural inhabitants, regardless of class or caste, and which has been successful particularly in those areas where a large cross section of rural society has felt dominated by the Maharashtrian ideological, political, cultural, and economic authority of the Maratha heartland to their west. In order for us to begin to understand how the idiom of Bali functions in the Sanghatana, and the reason that Bali is so closely identified with the subjectivity of oppression and disempowerment, we must examine the myth, and its intersection with Maharashtrian culture and society, much more deeply.

Bali's realm and his fall from rule

The myth of Bali's downfall is well known to Maharashtrians, and is easily recounted by informants from every caste and community with whom I have spoken about Bali. As recorded in the classical Puranic texts, the myth goes like this:[7]

> After Bali defeated the gods and extended his reign across the land, he asked his grandfather, the great demon Prahlada, for advice on how to govern the new realm. Prahlada replied: "Only virtue will always win. Rule the kingdom without deviating from virtue." Bali ruled his kingdom according to this advice, and he became famous throughout the three worlds.
>
> The subjects of the realm were happy and comfortable under Bali's rule—all, that is, except the gods and the brahmins, who felt they were denied the privileges they deserved. One by one, the gods began to leave the realm. Finally, Lord Indra, the king of the gods, approached Lord Vishnu and represented their grievances. Vishnu told them thus: "Bali is devoted to me. Still, to redress your grievances, I shall put him in his place."
>
> One day, as Bali was performing sacrifices to Vishnu on the bank of the river Narmada, he was approached by Vishnu himself—but in the incarnation of a dwarf brahmin hermit named Vamana, seeking alms. Bali was pleased with Vamana, whom he did not recognize as Vishnu, and offered to give him riches. But the hermit

rejected Bali's offers, humbly requesting instead that he be given only three steps of ground on which to meditate and offer sacrifices. Bali eagerly granted his wish, and Vamana proceeded to measure the ground with his steps. As he did so, however, he simultaneously began to expand to the stature of a giant. Fearful, Bali's subjects began to attack Vamana with anything they could lay their hands on, but still he continued to grow, becoming an immense being. With his very first step, he measured the whole earth. With the second step, he took the whole of heaven. With nowhere remaining to take the third step, he then asked Bali how he would be able to keep the promise that he had made. Bali, recognizing now that he was in the presence of Vishnu, replied that he had only his physical body remaining that he could call his own, and that the god might step on that in order to complete what had been promised. Thus Vishnu placed his foot on the head of Bali and pushed him down to Patal, where Bali was made to inhabit the nether world.

This central mythic event, for which Bali is best known, is the core of the Puranic version of a story called the *trivikrama*, the three steps.[8] It is a story told throughout the state—and it is also the key source of the Bali imagery used by leaders and participants in the context of the Shetkari Sanghatana. Although every oral account of the trivikrama that I have heard is distinguished by its own digressions and subtleties, most accounts embody the same core elements, what South Asian religion scholar Paula Richman (1991) has called the "skeleton" of a myth. This is exemplified in the following version told by an informant we will call Manjula, the fifty-one-year-old matriarch of a small landowning, kunbi household, in a village of Buldhana district. In our conversation, I had asked Manjula to explain the reason that so many rural Maharashtrians so often speak of the restoration of Bali's throne:

> **Author:** And why do you say this, that "Bali will return and all evil will be banished"?
>
> **Manjula:** Because Bali was a good king. He was good to the shetkaris, and his goodness was known by all the gods. The gods and the people respected him.
>
> **Author:** But then why was he deposed?
>
> **Manjula:** Because Indra grew jealous. Bali's kingdom was becoming so large that there was no place left for the gods. Indra called all the gods together...Shiva, Vishnu, Ganapati,[9] all of them....They tried to wage a war against Bali and the demons but they could not succeed. They tried again, and they still could not succeed. So Indra sent Vamana, who was really Vishnu but appearing as a very small man...a brahmin holy man, with no clothing or food. He knew that Bali was famous for being generous and, in fact, when Vamana came to Amaravati, Bali said "What can I give you, oh holy man?" Vamana was very clever. He said "I just want to sit and meditate. For that, I need only as much land as I can cover with three paces." And that is how he took away the kingdom...and the life of the shetkaris has gone from worse to worse ever since.
>
> **Author:** But the kingdom was so small?
>
> **Manjula:** No no, it was bigger than all of Maharashtra...bigger than America even. As soon as Vamana began to take his first step, he grew to a giant size and he stepped across the entire kingdom. He kept growing... enormous!...and the second step covered the

entire world, everything that there is. Then he
said to Bali "You promised three steps of land.
What am I to do?" Bali said "In that case, there is
nowhere else to put your foot but here" [tapping
the top of her head]. So Vishnu stepped on Bali's
head, and crushed him into Patal.

The major outlines of Manjula's story—including the three-
step deposition, the key characters involved (Bali, the brahmin
Vamana, Vishnu, Indra), and even the assertion that Bali was a
good king under whose administration most of his subjects
thrived—are all consistent with Puranic sources.[10] However,
many of the details in Manjula's version are idiosyncratic in
relation to the classical textual account—they clearly situate the
myth in agrarian and Maharashtrian contexts of meaning and
situate the teller in identity with a community of "shetkaris."

This suggests that the deep meaning of the myth does not
reside in the skeletal framework alone, but also in the narrative
detail that individuals and groups attach to the framework. These
are the nuances in the myth that reflect the beholder's broader
world of cultural understandings, lived experience, and terrains
of identity.[11]

Bali's realm and shetkari selfhood
In the context of the Sanghatana, this broader world of rural
experiences and identities is generally represented by leaders of
the movement as a shared "shetkari" experience—a composite
rural experience that, despite variations across its differing
constituents, is fundamentally defined by the experiential divide
between "India" and "Bharat." Sanghatana leaders invoking the
demon king draw a metaphoric parallel between the fate of Bali's
kingdom and this idealized, composite rural experience. Thus
Bali—like so many other Maharashtrian signs deployed at the
central level of the movement—is a discursive device that
functions to signify unity within the movement.[12]

Despite the wide range of interpretive possibilities, the Bali
myth is very effective as a signifier of this composite experience.

Its effectiveness stems from the myth's deep association with rural Maharashtrian culture and agrarian identity, and shared acceptance of the myth's basic skeleton in which a virtuous and successful kingdom is opposed and destroyed by an outside authority. At the shared skeletal level of the myth, Bali's conquest is not one that was achieved through superior force or even superior virtue—but, rather, through the deceptiveness of the victor and the generosity and honorability of the vanquished. This makes the Bali myth tremendously amenable to interpretations that comfortably correspond with the Sanghatana's depiction of all rural inhabitants as victims of an authority (India and the black British) that has impinged on their livelihood and expropriated their wealth through tricks such as the "negative subsidy."

As discussed in earlier chapters, these dominant themes within the Shetkari Sanghatana *and* within the myth of Bali's downfall are, in many ways, genuinely reflective of the lived experience of a very broad range of subjects within the Sanghatana's participant community. They are also themes that are common in rural Maharashtrian tales—such as tales in which agriculturalists or village servants are punished by gods and social superiors for their failure to do their duty of sharing their produce and labor.[13] For many in the agricultural community, therefore, the trivikrama myth mirrors substantial aspects of their subjectivity in deeply felt ways, regardless of their social position within rural society.

Nonetheless, the experiences and meanings signified in calls for the restoration of Bali's realm are complex, and they are open to other interpretations by each listening or invoking subject. For the rural shudra and Dalit communities—those members of village Maharashtra who are disproportionately landless, marginal landholders, or holders of the poorest quality fields (despite their size)—the trivikrama myth is also inherently a story of upper-caste authority and domination. In rural Maharashtra, caste hierarchy continues to parallel the hierarchical structure of socioeconomic power in many fundamental ways—and caste hierarchy itself is largely reinforced by upper castes through reference to Puranic (that is, brahminical, Sanskritic) ideology and practice. Thus, in order to better appreciate what Bali 'also' signifies for these least empowered segments of rural Maharashtra, it will be important to contextualize the trivikrama myth within a broader ideological context of brahminical Hinduism.

The "war" of brahminical gods against the gods' Others
The literary origins of the Bali myth can be traced to ancient Vedic and Puranic classical Sanskrit sources, but only in the later Puranic literature do we find the Bali myth as it is recognizable today.[14] This transition from the earliest Vedic texts to the later Puranic myths took place over a period of approximately one thousand years and corresponds with a number of important changes in South Asian society of the time. The evolution of theology and narrative styles during the period reflects a changing social landscape, marked by increasing assimilation between the Vedic Aryans and the non-Aryan indigenous cultures with which Vedic society had become intertwined. The transition is deeply evident in the "orthodox" Sanskritic Hinduism of today—or what anthropologist C. J. Fuller (1992) refers to as "brahminical standard Hinduism"—and is important to our understanding of the range of social meanings that may be implicated in the Puranic Bali myth.

One prominent transformation evident in the transition from the older to the newer literature is the centering of a new pantheon—most notably the trinity of Brahma, Vishnu and Shiva—and the apparent absorption within these gods of divine characteristics that are more representative of indigenous non-Aryan deities than of the earlier gods of the Vedas (Dimmitt and van Buitenen 1978; Kosambi 1994 [1962]).[15] It is probably correct to say, as suggested by Puranic scholars Cornelia Dimmitt and J. A. B. van Buitenen (1978), that this synthetic evolution from the "narrow orthodox brahminism" of the Vedas to the "popular…all-inclusive Hindu tradition" of the Puranas may have helped to promote a broad acceptance of Hindu self-identity across diverse regions and social groups (12–13). This adoption or co-optation of non-Aryan ideas and practices, however, must have also concurrently served to further consolidate brahminical claims to authority across broader terrains of popular practice, by enveloping greater expanses of the theological world within the ambit of brahminically sanctioned Hinduism (just as the three steps of the mythical brahmin Vamana encompassed ever greater expanses of "everything that there is"). Moreover, the Puranic literature supported these claims, in large part, through new forms of exclusion embodied in the representation of the gods' Others, the demons (cf. Jaiswal 1967; Kosambi 1994 [1962]). In this respect, two more transformations are crucially important.

One is that, while the gods of brahminical Hinduism were increasingly indigenized, so also were the demons. Although many popular non-Aryan religious beliefs and practices were amalgamated within the Puranic deities and their various avatars, other beliefs and practices that had not been legitimized under the new ideology were either absorbed into the characters and personalities of the demons, or crafted into new demons themselves.[16] As the renowned Maharashtrian archaeologist D. D. Kosambi (1994 [1962]) has noted, with regard to pre-Aryan village deities after the brahminization of Goa: "...their deities, where unabsorbed by the Brahmanic synthesis, have been converted to cacodemons, generally known as devchar [demons] but still worshipped by the lower castes" (166).

Thus, the "brahminic synthesis" represented an embrace of many extant popular spiritual ideas and practices, but also a rejection of others. The demons came to embody qualities resembling actual non-Aryan deities that were (and still are) venerated by non-brahmin castes.

Another important transformation is in the very meaning of the demons themselves. We can see this in the literary evolution of the most common and aggregate term for demons: *asura*.[17] In the earliest Vedic texts, the use of asura is somewhat comparable to the English word "lord." It was a title or rank that could be applied to lesser gods or to human rulers, with no inherent indication of whether they were friend or foe. Some asuras were enemies of the gods, or of humankind, and others were not. By the later Vedas, however, and very clearly by the time of the Puranas, demons came to represent a class of beings (again, with characteristics that are variously both human and supernatural) who are depicted as anti-gods—predominantly evil in character, mortal enemies of the true deities, and a terror to human beings (Hale 1999).[18]

All of this suggests that demons such as Bali, who are depicted in the Puranas as challengers of the gods, very likely were intended by the authors of the asura myths as metaphoric vessels, or generalized signifiers, of non-brahminical beliefs and practices of deity worship that were perceived as a threat to the emergent ideology. Thus, with specific regard to Bali, religion scholar D. D. Shulman (1980) hypothesizes that Bali represents a "big tradition" narrative on the moral downgrading of village patterns of worship. Indian historian Suvira Jaiswal (1967) takes this a step further, suggesting that the Bali myth "hints at the suppression of a [an existing, indigenous] cult of Bali" (124). Jaiswal, citing evidence from ancient texts containing instructions for making a cult image of king Bali, suggests that Bali was an established deity, with an existing cult following, prior to his representation as a demonic Other in Sanskritic myth.

Whether or not an actual cult of a Bali deity ever existed, or what may have been the social and geographic extent of his significance, is unknown. What is clear is that Bali signifies characteristics of many non-brahminical village deities that were no doubt deeply meaningful in the past, as they are today. Bali, thus, is metaphoric of a wide array of village spiritual ideas and practices—and the myth itself clearly treats Bali as metaphoric in these terms. Puranic narrations of the trivikrama are a story not only of the conquest of the demon Bali, but also of the expansion of the brahmin theocratic social order over the village social structures and antithetical worldviews with which it was in competition. Each of the three steps in the myth can be read as a completion of this conquest. After revealing that the brahmin mendicant is, in essence, Vishnu himself with authority over all that the brahminical pantheon is entitled to command, the god proceeds to delineate the exact (and total) terrains of that control. In the first step, Vishnu encompasses the physical world. In the second, the rest of the physical universe and the spiritual domain of "heaven" (where, in the Puranic accounts, Vishnu receives the blessing of the earlier Vedic god Indra, that he may continue to extend his domain). The third and final step signifies the defeat of all contenders—particularly the unincorporated folk deities and demons venerated by the non-brahmin populace—and the deposition of these to a nether world far removed from the domain of either humans or gods.

The Bali myth is just one of many that abound in Hindu folklore and Puranic literature concerning confrontations between the gods and the demons. Narratives usually begin with a set-up such as "the demon X was threatening the world" or "X was a threat to the supremacy of the gods." If we consider these battle myths as even slight hints about actual confrontations between the vanguard of an emergent brahminical social order and the local orders already in place, then the conflict and its social upheaval was as dramatic as it was arduous. In the texts, battles are depicted as epoch events, often without clear victors on either side. Demons and their armies were often victorious in isolated clashes, only to meet their eventual downfall in a final spectacular bloodletting or in a deceptive trap laid by the jealous gods. In Dimmitt and van Buitenen's translation of another episode in the contest—the Puranic account of Bali's first royal assault on the gods, staged just after his coronation as king of the demons—the terrifying imagery of the battle is striking:

> When he heard about [Bali's] demon army, Sakra [Indra], leader of the gods, said, "Let's go to war with these demons who have assembled their armies!" So speaking, the mighty lord, king of the gods, swiftly mounted his own vehicles and sallied forth for battle…The clash between those two armies on that mountain was horrible, O seer, like the ancient war between the monkeys

and the elephants. Battle-dust churned up before
the fray was ruddy, like a cloud in the sky
reddened by the sun at twilight...Nothing could
be made out in the gloom, but on all sides people
cried out "Cut!" and "Slash!" without pause. A
ghastly, deathly river of blood pouring from both
demons and gods began to settle that cloud of
dust (1978, 300).[19]

Scholars are uncertain of the extent to which narratives of
conquest in these texts may reflect the real character of any
historic clash between societies and ideologies. While there is
some evidence that Aryan migrants into South Asia may have
marauded over cities and villages, for the most part the
archaeological record suggests a drawn out process of gradual
institutional and cultural assimilation between the Aryans and
the non-Aryan indigenous communities (Wolpert 1993).
Nonetheless, the instructive intent of these battle narratives for
their audience is clear: acknowledging the supremacy of the
brahminical gods and brahminical authority, was essential for the
establishment of an ordered society. For example, recounting the
gods' sacking of the city of Tripura, whose inhabitants had been
led astray by the demon *Maya* (whose name means illusion or
falsehood), the authors of the *Matsya Purana* leave no question
about the lessons of the new ideology:

So the demons who lived in Tripura were ruined
by fate. They deserted truth and virtue and did
instead what was forbidden. They despised the
holy brahmins, failed to revere the gods, ignored
their teachers, and even began to hate each
other...[They] mocked their own duty...They
contended with the brahmins, thus following
their own will like the elder gods before
them...The whole world was devastated by those
evil enemies of the immortals, like crops overrun

by a swarm of locusts (Dimmitt and van Buitenen
1978, 191–92).[20]

Of course, this is myth—and, as myth, we do not want to read
too literally or conflate it with an accurate representation of
historic occurrence. Ancient myths are often grounded in real
events, but they are usually stories that have been crafted,
interpreted, and re-crafted over centuries of successive oral
accounts in order to serve not only evolving discourses, but also
listeners' demands for drama and entertainment. At the same
time, however, myths are not meaningless tales or dead stories.
One way to think of them is as what religion scholar Wendy
Doniger O'Flaherty (1988, 25) has called "operative fictions":
stories that cannot be told or heard without an inherent act of
interpretation that roots the myth, in one way or another, to the
beholder's own understanding and experience of the world.
Moreover, myths not only *reflect* the beholder's experience, they
can also *structure* the beholder's experience of the world in ways
that are informed by his or her interpretation of myths. Thus, from
an ethnographic perspective, the way a myth is understood and
experienced by listeners and tellers is far more important than the
question of the myth's original truth.[21] And here it is worth
recalling that my informant Manjula's account of the Bali myth
did not fail to overlook the successive battles between the demons
and the gods, or the social guise of Bali's conqueror: "Indra called
all the gods together…They tried to wage a war against Bali and
the demons but they could not succeed. They tried again, and they
still could not succeed. So Indra sent Vamana…appearing as a
very small man…a brahmin holy man…"

Myths of demon conquest *do* live in rural Maharashtrian
cultural experience, and are far from dead stories. It is crucially
important that, although the latest contributions to the Puranas
probably date back some one thousand years, they continue to
provide not only the critical theological reference points that are
today invoked in most conceptions of "orthodox" brahminical
Hinduism, but also the dominant ideological lens through which
most heterodox popular Hinduisms are perceived and evaluated

by brahminically influenced society—particularly by the urban establishment and the Sanskritized rural elite (Fuller 1992). This is deeply significant considering that in rural Maharashtra today the worship of heterodox local deities—many of them specific to different caste groups, villages, or areas of the state—continues to be, in many ways, more important in the fabric of everyday life than the acknowledgement of brahminical Puranic deities (cf. Feldhaus 1995; Karve 1968). Hence, to consider these mythic narratives and the lessons they tell simply as stories and entertaining diversions that have long since lost their significance to lived reality, is to overlook the ways in which these are imbricated with the experience of daily village life.

The lived context of Puranic worldview in Maharashtra

It is important to recognize that in Maharashtra today, as at other times in the historical record of Hinduism, spiritual worldview and practice is characterized more by a plurality of ideological norms than by a strict social continuity of rules or ideas. While there are certainly deep cultural similarities of practice within the Hindu world,[22] local Hinduisms, particularly as practiced by the least Sanskritized rural communities, differ substantially in their understandings of the supernatural world and in their assignment of names, forms, and characteristics to supernatural beings. This is why most modern scholars of Hinduism conceptualize the religion as an ortho*praxy*, rather than attempting to identify it with any singular and extent ortho*doxy*.[23] Maharashtra is exemplary of this complexity. When we think about this broader world of folklore and folk practice in Maharashtra, it is important to recall that the Marathi-speaking regions constitute a state of tremendous diversity, rooted in distinctive histories, varying degrees of integration with the Maratha heartland, and ongoing caste-claims or competitions for social status.

In the face of this diversity of Hindu thought, struggles to define or resist the nature and meaning of "orthodox" Hinduism have been as important to the meaning of worldview and practice in the past as they are today. Much of the reason for this is that the distribution and dissemination of heterodox ideas and practices is far from random: borderlands of competing doxies, though rarely precise or immovable, have always existed between different regions and cultural terrains, between different caste groups and lineages, and between town and country. These often fuzzy boundaries—much like the boundaries of caste groups and other lines of social, political, and economic status—have been repeatedly asserted or negated in the contexts of major and minor migrations, social upheavals, and changes in governance. They have also been (and still are) asserted or negated in the routine social exchanges of daily life, whenever the boundaries of conceptualized spatial, cultural, or socioeconomic realms are encountered and negotiated. The struggle to define "orthodoxy" is, in the final analysis, a struggle of, and for, relative power. This is a struggle that is embedded not only in the trivikrama myth itself, but also in much of Maharashtrian identity discourse in general.

Social hierarchy and daily discourses of spiritual Otherness
One of the ways that this is embedded in daily discourses on the terrains of identity and authority can be seen in upper-caste commentaries on the spiritual practices of rural shudra, Dalit, and adivasi communities. As Fuller (1992) states with regard to popular Hinduism throughout India:

> ...the Brahminical standard of Sanskritic Hinduism is neither monolithic nor unchallenged by alternatives. Nonetheless, it represents the single most important evaluative norm within Hinduism. By reference to it, higher-status groups tend to regard their own beliefs and practices as superior to those presumed to belong to lower-status groups (27).

The operative nature of this evaluative norm is readily discernable in Maharashtra.

Photo 14. Vishnu saving the world from Bali, as depicted in a sixth-century cave carving at Badami, just south of Maharashtra in northern Karnataka. Several episodes of the myth are depicted in the scene: Vamana approaches Bali and his royal court (right side of scene, on the ground beneath Vishnu's raised leg); a giant Vishnu takes his third step on Bali (who is shown upside-down directly below Vishnu's raised foot); Bali accepts his fate and clings to Vishnu's leg in devotion (left side of scene).

Photo 15. Detail of Photo 14 (opposite).

As mentioned in chapter 2, the prominent Maharashtrian cult of Khandoba, embraced particularly by members of the relatively low-status Dhangar caste, is widely denigrated for espousing a form of bhakti (*sakama bhakti*) in which practitioners offer devotion to god with the expectation of getting something in return. Upper caste members typically reject this form of devotion (in public at least, if not always in personal practice) as corrupt. Interestingly, this is a form of human-deity relations that is much more common in folk practice in general, in which individuals frequently appeal to local deities, demons and spirits for the fulfillment of favors and demands such as advantageous weather, cure of an illness, or resolution of personal difficulties.

The social logic for elite refutation of such forms of human-deity relations is clear. From a brahmin ideological perspective, they substantially contradict brahminical versions of "orthodoxy" and thus also the bases for brahminical ideological authority. Moreover, in actual practice, these forms of worship negate the necessity of brahminical priestly intervention—just as is the case with other forms of direct worship, such as the bhakti of the Varkari Sampradaya. But this form of worship could also be seen as a challenge to any group of village socioeconomic elites, brahmin or otherwise. Because such "giving deities"—whether Khandoba, or any other local deities and village demons—are popularly identified with material success and power (Shulman 1980), worshipping them partially sidelines human overlords who seek respect and authority as providers of livelihood. This affront to power may help to explain, in part, why most of these non-Puranic spiritual practices of the village lower castes are conducted in the domestic sphere, or in peripheral spaces outside of the central village—such as in a field, near a river, or in a sacred grove—rather than in the village center.

Photo 16. Vamana steps on Bali's head in a Shetkari Sanghatana poster. The text warns that wherever the kingdom of Bali appears it is being crushed by ideologies of communalism and divisiveness.

Of course, when upper caste members speak of folk practices in negative terms they do not normally talk in terms of threats to their socioeconomic superiority. Rather, upper-caste villagers typically denigrate lower-caste spiritual practices by calling them superstitious, unsophisticated, peculiar, or just simply foolish. This perspective is clear, for example, in the comments of one upper-caste informant I will call Vivek, a member of a large landholding family of the locally elite Lingayat community[24] in a village in Marathwada. In Vivek's words, these beliefs and practices of his village's lower castes "… are for the simple people who don't know any better. They may hear something behind a tree, and straightaway it is a *bhut* [ghost]; or someone is sick, and just like that it is because of *devi* [a folk goddess], and some ritual or another must be done to make her happy.[25] If they are in debt, or there is no rain, they may turn to some *rakshasa* [demon] for help."

This is not to say that such practices and beliefs are, in the norm, actively prohibited at the village level. In most routine village contexts, expressed attitudes toward these less Sanskritic folk practices are highly contextual. In some situations, upper-caste members may refer to these practices and beliefs as examples of the irrationality of lower castes—maybe even as evidence of their inability to hold positions of responsibility or to wrest their way out of adverse financial circumstances. In other contexts, these same practices may be patronized by elite villagers who find themselves in extraordinary circumstances of their own, requiring alternative ("dangerous" or "impure") supernatural interventions. Thus, in some respects, heterodox practices and beliefs are also valued by the upper castes as the special culture and knowledge of their village Others—but the implicit understanding seems to be that these should remain on the periphery of village social and cultural life.

This dislocation of lower-caste beliefs and practices to the periphery is metaphorically represented in public interactions with myth. As elsewhere in India, religious culture in Maharashtra is rich with stories in which archetypical demons are manifested as marauding Godzilla-like characters that are ultimately vanquished by one or more heroic deities. These myths are not only well known, they are also deeply embedded in the cultural and spiritual life of village Maharashtra.[26] Some of these narratives are dramatically enacted in major Hindu festivals celebrated during the course of the calendar year—reasserting each festival season the primacy of the brahminically-orthodox pantheon and the subjugation of the village deities and demons that most members of the rural communities, in other contexts, may highly revere. These festivals create occasions not only for dramatizations of conquest, but also dramatizations of resistance—and in Maharashtra, much of this conquest and resistance is couched in the symbolism of the demon Bali.

Conquest and the return of Bali in the festival of Navaratri

The prominent annual festival of *Navaratri*, which is held in Maharashtrian villages (as elsewhere) during the first nine nights of the bright lunar fortnight of *Ashvina* (September–October), entails a celebration of the defeat of the buffalo-headed demon *Mahishasura* by the goddess Durga.[27] This period in the calendar overlays another major event in Puranic myth—Lord Ram's killing of the demon *Ravana*, described in the epic Ramayana.[28] As in many other parts of India, village celebrations on the final day of Navaratri (the tenth day, known as Dasara) are increasingly characterized by the celebratory burning of Ravana effigies in the village center or on other public grounds. This is often accompanied by the lighting of firecrackers and other festival fanfare. Most introductory texts on Hinduism, as well as many upper-caste villagers, describe Navaratri and Dasara as a celebration of the victory of "good" over "evil." In its actual village practices, however, this festival complex reveals many deeper and far more complicated dimensions of meaning—and these meanings are deeply intertwined with local assertions and counter-assertions on the legitimacy of spiritual, political, and economic inequality.

At the opening of this chapter, I described one such practice, as conducted by a Deshasta brahmin husband and wife whose ancestors were the ruling elite in a large village that I will call Gordpani in Ahmadnagar district. Ashok's symbolic disembowelment of Bali with the tip of a kitchen knife is a microcosm of other more public celebrations of demon conquest, and vividly ties together the mythological world of demon-killing and the village world of socioeconomic hierarchy. This ritual is conducted on the demon-killing day of Dasara.

On the same day the brahmins of Gordpani are symbolically extracting agricultural wealth from Bali—and villagers and townspeople across the state are burning effigies of a demon— ordinary agriculturalists throughout Maharashtra are engaged in the culmination of a ten-day ritual that markedly contrasts with these activities. This ritual, called *Ghatstapana*, begins on the first day of Navaratri and continues through to the final day, each step of the way symbolically asserting the economic and spiritual autonomy of individual cultivators. Like the Bali-stabbing ritual, Ghatstapana begins within the household. It involves no mediation by brahmins or any other authorities.

As I have observed the practice of Ghatstapana in a small, brahmin-less village I will call Neemgaon, just a few miles from Gordpani, it proceeds as follows. On the first day of Navaratri, the non-brahmin agriculturalists and other low-caste villagers make a dish-shaped bed of leaves. Upon this bed, they place some soil from their own land (or, if they have no cultivable land, soil from the immediate area surrounding their home), and they seed this soil with several different grains. Finally, a *ghat* (a small clay vessel) filled with water is sunk into the center of the soil, and the whole fixture is installed in the small household *devaghur* (godhouse), nestled among the numerous images of both Puranic and local gods. All of this is usually done by an older woman of the household. Over the course of the nine days of Navaratri, the soil is watered as needed in order for the seeds to sprout. On the tenth day—the culminating day in the celebration of "the victory of good over evil"—the ghat and the sprouted soil on the bed of leaves are tossed or buried in the family's field while participants utter the words "Let all evil go away, and the kingdom of Bali be restored."

There are a number of interesting ways that we could interpret Ghatstapana. The soil and ghat, placed in the godhouse, suggest a veneration of the practitioner's own land and access to water—or a request from the gods for the acquisition of these. The burial of the sprouted seeds and soil in the field suggest a rite of fertility—or perhaps even the anticipated birth and return of Bali, rising up from his exile in Patal. Most importantly, however, the call for Bali's return, in the context of the demon-killing festival, suggests deep differences of opinion on the social meaning of the demon himself and all that he signifies.

Ghatstapana is practiced throughout Maharashtra, but there is quite a range of other village practices that also involve Bali on the day of Dasara. As one informant from Sangli district described events in her village:

> After the ghat is buried, the shetkaris go around exchanging leaves of the apte tree, as if they were gold. We make little earthen dolls of Bali and place them outside the home...Women approach neighbors saying "Let all bad things go away, the kingdom of Bali will come." But, what happens is that, when a brahmin comes along, he takes a long stick and he pokes it into the belly of that Bali doll as he walks past. I don't know how long this has been going on, but that's how it is.

Appropriating Bali's kingdom in the festival of Diwali
The other major festival in the same season is Diwali, which follows on the heels of Navaratri and is an extension of the narration of Rama's victory over the demon Ravana. Diwali lasts four to five days[29] and is often described as "the festival of lights." It is nominally dedicated to the worship of the Puranic goddess Lakshmi, in her aspect as Wealth—but it is also a continuing celebration of the homeward march of the kshatriya god-man Rama with his virtuous wife Sita (whom he had just liberated

from the clutches of Ravana), and his return from exile to reclaim his rightful throne as king. In urban Maharashtra, as in other parts of India, a prominent and popular aspect of the festival is the lighting of bright candles and electric lamps—a practice that is variously described as a welcome to Lakshmi, a symbolic lighting of the route home for Rama and Sita, or a celebration of one's financial success in the preceding year. In rural Maharashtra, lighting lamps for Diwali is much less common though it is gaining in popularity as a borrowed practice from urban culture. Rural areas have other, more typically rural practices. The final day of Diwali coincides with a day of special significance to Maharashtrian lower castes—a day called *Balipratipada*, the commemorative day of Bali's defeat. This day, according to older informants, has been significant in rural Maharashtra for much longer than the lighting of lamps, the public celebration of wealth, and other gala displays.

Balipratipada is commemorative of Bali in a number of different ways, each embodying different meanings for those who engage in the commemoration. In a number of villages, for example, informants have described that on Balipratipada brahmin or elite Maratha men perform a brief ritual of crushing a fruit, using a hand, a stick, or their foot. This, they say, is a symbol of Bali, being crushed on his day of defeat. In a village I will call Tapa, in Vidarbha region, brahmins and other upper-caste villagers engage in a ritual that is even more to the point. This was described to me by an informant I'll refer to as Satish, an educated brahmin social worker who returned to Tapa with plans to launch economic development projects:

> What I have seen in my village is this: During the two days at the end of the year, and then on the *padva* [the beginning of the Hindu new year], all the upper-caste villagers make an image of Bali out of cow dung, and install it by the door of the house. On the padva, then, they explode this image with firecrackers. Then the Dalits go from

door to door, saying "Bali will return," and beg
for Diwali alms from the upper castes.

Satish went on to offer his interpretation:

> As you know, the cultivators all call themselves
> Bali, as if it were the most honorable thing you
> could call another person...but in these villages,
> the Dalits identify with Bali even more.
> Balipratipada, you see, is an annual celebration of
> the upper castes' defeat of the lower. When the
> non-caste [Dalit] Hindus go begging on
> Balipratipada, it is as if they are the vanquished
> asking for the mercy of the victors. Bali is
> exploded...this is total defeat. Then the low
> castes go around the village and say "Now we are
> defeated, now we are beggars. Please show some
> mercy on us." And they receive sweets and gifts
> and bonuses, and so forth.

Although this focus on Bali is deeply Maharashtrian, the
receipt of gifts, sweets, or wage bonuses to which Satish referred
is a key convention of Diwali as celebrated in other parts of India
as well. While village possessors of capital—traders, processors,
shopkeepers, and sometimes large, labor-employing
landowners—view Diwali particularly as a festival for the
worship of wealth (that is, the goddess Lakshmi, but also their
own acquired wealth), and treat Diwali as the end of their
"business" fiscal year, agricultural laborers, toilers, and rural
workers view Diwali largely as an opportunity to claim an annual
bonus or small gift from their employers. Satish's comments on
the subjective significance of this are echoed by the comments of
an informant we'll call Balaram, a thirty year-old Mahar Dalit,
from a marginal landholding family in Beed district:

> When the laboring people in the village go out to
> visit the wealthy families, asking for some bonus
> or something to eat, they do so with an
> expectation that the strong, the ruling section of

> society should show some generosity to the
> weak. Yes, this is a festival season, but it is the one
> time of the year the poor are most conscious of
> their own weakness.

While all of these practices suggest deeply rooted contests over the meaning of Bali and the legitimacy of all that Bali signifies within the village community, it is important to keep in mind that these contours of meaning do not represent a strict spiritual (or material) dichotomy between Puranically oriented elites and heterodox subalterns. These contests may be more or less prominent in different village settings, and the terrains of identity with Bali (or other beings antagonistic to "orthodox" brahminism) may shift from context to context. Village elites do not practice strict abstinence from the supernatural worldview and practices of the rural masses; the masses, in turn, do not abstain from engagement with popular Puranic Hinduism and its public displays.

In fact, for most lower-caste Hindus, folk practices associated with local deities, demons, and spirits fit quite comfortably alongside the dominant Puranic worldview. Shudras, Dalits, and even many adivasi communities commonly participate in the major Puranic festivals, and they visit the village temples of Puranic gods such as Maruti[30] or Shiva. Moreover, they typically have images of the major gods and goddesses in their household shrines alongside objects or images representing more subaltern deities and demons, and householders can usually narrate stories that establish the identity of the latter as incarnations of the Puranic deities—similar to the Sanskritization of Vithoba and Khandoba. Thus, within most sections of the village's Hindu community (and even among most Ambedkarite Buddhist Mahars) Puranic culture and practice is rarely rejected wholesale. For the most part, major elements of Puranic norms are embraced by the lower castes, and are as deeply entwined in daily life as is the case for village upper castes. So, while in some contexts subjects may interpret Puranic myths of confrontations with non-Puranic spiritual forms as narratives of social and ideological domination, we cannot assume that such interpretations are always in the forefront of subjects' consciousness.

At the same time, it is clear that in some contexts subjects do consciously recognize the contrast between that which is approved by ideological power and that which is not. Very few of my kunbi and Dalit informants were oblivious to the subtle (and not so subtle) ways that festivals such as Navaratri and Diwali seem to reinforce local social hierarchies and continue the discursive denigration of many of their own spiritual practices. This is something they feel in other ways throughout the year.

Projecting Bali into the social center—the agrarian Pola festival
While some of the upper-caste practices that symbolically reenact the conquest of Bali are conducted in public spaces, lower-caste practices of Bali veneration are far more understated and generally confined to the domestic sphere, the near environs of the home, or the family field. This confinement of Bali is socially reinforced not only by the dominant narrative on demons in general but also by practices described above, such as when an upper-caste passerby crushes an image of Bali hung outside a lower-caste home. Thus, most of the practices that assert the validity of Bali are extremely confined, such as installing the ghat—a symbol of Bali's return—in the family godhouse amongst the brahminically-approved deities.

Nonetheless, despite reinforcements of social confinement, we have also seen ways in which shudra and Dalit villagers stealthily project Bali into public space. Hanging Bali images outside the home and burying the ghat in the family field both extend Bali's realm beyond the interior confines of the house. Declaring in public that "Bali will return," and publicly addressing fellow agriculturalists and laborers as "Bali," verbally asserts the validity of Bali in village lanes and public centers. One agrarian practice that we have not yet discussed achieves this more than any other—the veneration of bulls and bullocks in the festival of *Pola*. This practice is widely described by lower-caste villagers as a form of *puja* (worship) for Bali.

Photo 17. *A village woman prepares a ghat on the first day of Navaratri and installs it in the household god house.*

Photo 18. *Ten days later. The ghat has sprouted and is ready to be tossed into the family field with the words "Let all evil pass and the kingdom of Bali return."*

Pola (also called *beilpola*) is a rural festival practiced by village agrarian and laboring castes in the lunar month of *Shravan* (July to August). Nominally, it is a day of compensatory veneration for the bulls and bullocks that, on every other day of the year, labor and toil for their masters' livelihood.[31] On Pola, these "servants" are treated, literally, like kings. They are decorated in rich colors with paint and powders. Sometimes they are garlanded with marigold flowers, or their horns painted gold. They are treated to a day without labor, fed sweets and the finest fodder, and—most importantly—paraded through the village center, for all to see their grandeur, as if kings on a royal tour.

On Pola the bull or bullock enjoys a day of social inversion in which the servant becomes the king. However, there is more going on: the animal is treated not just as a king, but also as a representation of the king of demons, Bali. Thus, on one level it is not merely a bull or bullock that enjoys the social inversion and is paraded around the village for all to appreciate, it is Bali. Two things are important about the potential significance of this act. First, the bull or bullock (*balivard* or *balivahan*) is generally symbolic of Bali—not merely on Pola, but also on every other day of the year when the animal is a servant and is, by its very name, a "helper of Bali." Second, village agriculturalists and lower castes are also symbolic of Bali (and vice versa). Hence, the veneration of the bull or bullock is simultaneously a veneration of Bali, which in turn is a veneration of the lower-caste and agrarian self. In this we see a transference of the object of veneration that is not unlike that which we have discussed in the Varkari Sampradaya, in which the god Vithoba is embodied in the medieval saints, who are in turn embodied in the Varkari pilgrim. This explains why many informants have described their veneration of bulls and bullocks on Pola not only as Bali puja, but also as a sort of self-puja.

Pola is a relatively subtle practice—far less overt and assertive than the public burning of demon effigies, or even the squashing of a fruit under an upper-classman's foot—but these aspects of Pola suggest a widely felt validation of the demon victim of

Puranic conquest. As with other expressions of Bali veneration we have considered, Pola suggests that the validation of Bali expressed within the Shetkari Sanghatana is in many ways an expression of low-caste and agrarian subjectivity. Indeed many informants do interpret the symbolism of Bali as a challenge to the socioeconomic and cultural hierarchy in the village community. The more closely we look at the myth of Bali and its varied practices and implications in Maharashtrian society, the more we are able to recognize Bali Raja as a vehicle through which these material and cultural hierarchies are debated, challenged, or asserted whenever he is invoked. This is extremely significant when we consider that the most public and overt projections of Bali veneration into the social center of rural Maharashtrian life have been through the medium of the Shetkari Sanghatana.

Photo 19. A bullock is gaily decorated with bright colors for Pola. On this day the bullock is treated as a king and symbolically identified with Bali.

Bali veneration in the context of the movement

Bali is an important Maharashtrian figure, widely identified with the agrarian and laboring castes and popularly metaphoric of low-caste and Dalit aspirations for socioeconomic equality and opportunity. Bali is also an object of veneration, and serves as a cultural idiom of both conquest and resistance. In this respect, Bali is a metaphor for a non-orthodox social and spiritual worldview —the social and spiritual worldview of the least advantaged segments of rural society. Unlike other major supernatural figures of veneration, however, Bali has no widely recognized iconic form or visage of his own and no existing sectarian cult of worship. Jyotirao Phule and the Satyashodhak Samaj made a strong push to shift Bali veneration into the center of village life in the nineteenth century, but formal spiritual practice along Satyashodhak ideological and ritual lines is not prevalent today. What Bali *does* have in contemporary Maharashtra is the Shetkari Sanghatana.

The advancement of Bali as a major idiom of struggle within the Sanghatana community is a collective social act that vaults the demon king into the center of rural public life to an extent that has been unprecedented and, in many ways, socially prohibited in other contexts. But it is also a collective act that operates on multiple levels. While all rural inhabitants—including the village elite—may be able to identify with Bali as a signifier of rural aspirations for increased economic autonomy and freedom from the interventionist policies of the black British, many small-scale producers and laborers—predominantly low-caste and scheduled-caste villagers—also see in the Sanghatana's main-streaming of the heroic Bali image a vindication of their own resistance to the local structures of power and authority that shape their daily lives. These small producers and laborers have embraced and deployed the idiom of Bali within the Sanghatana community as a resource for projecting these oppositional meanings into daily discourse and practice. Emboldened and in-formed by their own engagement with the movement, and supported by the Sanghatana community at large because of the ambiguity that Bali encompasses, these segments of the community have influenced and promoted the emergence of new expressions and practices that validate the subaltern meaning of Bali.

Some of these new expressions represent new forms of village-level resistance to the dominant village atmosphere of Sanskritically aligned hierarchy and authority. Notable among these are the rejection of key aspects of the major Puranic festivals of Navaratri and Diwali, including:

Increased emphasis on Ghatstapana: In many villages where there is a high level of participation in the Sanghatana, informants describe seeing an increasingly public emphasis on Ghatstapana and on declarations by villagers that "Evil will pass, and the kingdom of Bali will return." In the interpretive context of the movement's "loud voice" ideology, calling for the return of Bali Raj is comparable to lauding the achievements and future success of the Shetkari Sanghatana. For the rural lower castes, however, it also embodies a validation of the self and of a worldview that is commonly denigrated by the upper castes.[32]

Boycotting Diwali: Another readily discernable trend within villages that have a high level of engagement with the Sanghatana is the outright boycott of pan-Indian elements of Diwali, such as the lighting of lamps and the ritualized veneration of wealth. All Sanghatana participants may be able to embrace this as a rejection of the encroachment of urban, cosmopolitan culture into the village. For laborers, smallholders, and indebted families, however, this can also be seen as a rejection of a conquest festival—as well as of festival practices that are easily associated with moneylenders, landlords, traders, and bankers who have made their wealth on the backs of others. Women in the Sanghatana also commonly view the boycott of Diwali as a rejection of patriarchal domination, as symbolized by Rama's treatment of his wife Sita.[33]

The boycott of Diwali began at a grassroots level in several villages, and was at the time of my fieldwork promoted by

Sanghatana leaders. As Rahul, the district-level leader in Marathwada quoted earlier, described it:

> Diwali was never really popular in our villages—but it has been becoming so with the new, younger generation. Many of the villages have made it clear that they want this trend to stop...but, in others, where it has become popular, we say to them "You are always asking for the return of the reign of Bali, right? That means there must have been such a king of the shetkaris, and the shetkaris must have prospered. So why do you celebrate a festival that is the victory celebration of your king's conquerors?"

Burning effigies of Vamana: In many villages, participants have taken the boycott of Diwali a step further. Crafting the occasion into an "anti-Diwali," sanghataks celebrate the festival season with a neoritualized burning of an effigy of Bali's conqueror, the brahmin dwarf Vamana who deceived him. Participants in the Bali Rajya village of Dorlapur have been boycotting Diwali and burning Vamana effigies for several years. Informants in other parts of Marathwada and Vidarbha (and to a lesser extent Western Maharashtra) have indicated that this practice is becoming increasingly popular. In Dorlapur's surrounding district, for example, one district-level organizer explained to me (in 1998) that the public burning of Vamana effigies was being practiced by assembled Sanghatana participants in "approximately forty villages in this district alone"—and this is but one of seven districts in the Marathwada region.

The burning of Vamana effigies is a deliberate response to the historic sidelining and maligning of Bali—but, as with the other expressions above, the meaning of the action can be interpreted in different ways by differently situated participants. On one level, the effigy of Vamana can represent the jealous and deceptive black British of India; on another level, Vamana (a brahmin) can represent an internal upper-caste or upper-class authority; and on yet another level, Vamana can represent a more generalized rural socioeconomic and spiritual hierarchy.

These public boycotts and displays of opposition not only project Bali into the public sphere, they also constitute moments in the creation and articulation of the Sanghatana's community idiom. In these expressions, Bali signifies far more than the Sanghatana's proposition of a fundamental divide between Bharat and India. He also signifies social tension and the potential for resolution between variously situated groups within the rural community itself. Through the involvement of smallholding and laboring participants in the Sanghatana, these nuanced layers of meaning are embodied in other expressions of Bali that we have seen throughout the movement, including:

Signifying Bali as the king of shetkaris: When Sanghatana leaders and participants represent Bali as the deposed king of shetkaris, they are overtly using the demon king as a metaphor for indigenous rural rule that is characterized by integrity, commitment to duty, generosity, and virtue. All of these qualities are ascribed to Bali in the Puranic myths. On a more protestant plane of interpretation, however, Bali is also a symbolic alternative to the Maharashtrian national narrative of indigenous rule under Shivaji. Notably, unlike Shivaji, Bali is neither Maratha nor kshatriya—and thus signifies the legitimacy of rule under a lower-caste (and less Sanskritized) king.

Establishing Bali Rajya villages: The declaration of participating villages as "Bali Rajya villages" signifies local aspirations for greater economic and political autonomy. It also signifies the solidarity—or the ideal of solidarity—of castes and individuals within the village and villages with the movement. Claiming such symbolic personal and village-wide identity with subjecthood under Bali's realm is part of a general rural claim to victimhood. At the same time, however, it is also unavoidably a symbolic claim of identity with those least empowered segments of the rural community who themselves most identify personally with Bali.

Calling for Bali Rajya Marathwada and Vidarbha: Just as declaring villages as "Bali Rajya gavs" aligns the idea of the oppressed village as a whole with the oppression of the poorest members of village society, calling for Marathwada and Vidarbha statehood under the rubric of Bali's realm highlights the ways in which these regions are dominated by political, economic, and cultural power from the west of the state. Relatively well-to-do cultivators—such as the cotton grower Rameshrao in Dorlapur—may view the call for a Bali Rajya state in just these terms. However, as the Mahar laborer Balu made clear, some members of the community view the "real" realm of Bali as the place in which the poorest of any village reside. From Balu's perspective, a true Bali Rajya state would address not only conflicts between India and Bharat—it would also prioritize the struggle for Dalit and low-caste equality. This meaning is embedded in Balu's own use of the idiom whenever he participates in the movement.

Representing Sharad Joshi as Bali: Finally, when participants represent Joshi, the leader of the movement, as Bali-like—often to the extent of calling him an incarnation of Bali—they signify equally complicated and nuanced layers of meaning. On one level, such representations suggest that Joshi embodies the Bali-like qualities of integrity, commitment to duty, generosity, and virtue that are widely perceived to be lacking in the current governmental order. On another level, however, it projects Joshi into the public realm as the embodiment of an agrarian and low-caste ideal. I will discuss this in some greater detail in the next section.

From a lower-caste and lower-class perspective, all of these uses of the Sanghatana idiom of Bali project the demon king, and all that he is capable of signifying, into the public sphere in exceptional ways. The depth and breadth of this projection are made possible by the scale of the movement, the diversity of its socioeconomic base, and the tremendous communication resources available to the movement's leaders. Each time Bali is invoked within the multicaste, multiclass movement—whether in the declarations and writings of leaders or in the routine interactions, public agitations, and fist-raising cheers of ordinary participants—all these layers of meaning are effectively lifted into the larger Sanghatana idiom and broadcast throughout rural Maharashtra to be interpreted, considered, and engaged with in different ways by a multiplicity of subjects.

Reconsidering Sharad Joshi as the demon king

Before concluding, I want to take a closer look at one final puzzle: the deification—or Bali-fication—of the movement's central leader. My experience with the activist named Sandeep, described in the first chapter, was not the only time I saw Sharad Joshi's image in a household godhouse, nor the only way that I would see Joshi venerated beyond the status of ordinary men. During my years of fieldwork, I often heard Sanghatana participants refer to him as a god or as an incarnation of Bali. I saw many images of him in framed photographs on participants' walls, garlanded and hanging high near the ceiling amongst the family ancestors; on Hindu wedding invitations, where auspicious images of the goddess Lakshmi or other deities would more conventionally be printed; on the dashboards of rural rickshaws and trucks, where drivers would more commonly install images of spiritual gurus and patron gods; and on handouts and flyers produced by participants at their own expense, often depicting Joshi as Bali himself. How can we make sense of this veneration? This particular puzzle becomes even more challenging when we consider that Sharad Joshi was, from the beginning of the Sanghatana movement, represented by urban observers as something of a fluke and an unlikely agrarian leader.

At first glance, Joshi may indeed seem like a paradoxical leadership figure for rural Maharashtra. As an intellectual, an urbanite, a brahmin, and a former bureaucrat, he seems to have none of the characteristics with which the rural masses would identify. Moreover, Joshi himself has done little to "ruralize" his image and even less to promote his own deification. Most agrarian leaders in India tout their relationship to the land, to a village, or to an agrarian caste, and package their public persona in the clothes and behaviors evocative of that identity. Sharad Joshi, in contrast, typically appears before villagers in crisp blue jeans, golf shirts, and tennis shoes.[34] He rebukes participants for treating him as anything more than a fellow activist.[35] In order to make sense of this seeming paradox, let us begin by considering some of the key bases by which existing and potential leaders are assessed in Maharashtrian public discourse.

Joshi and the cultural terms of legitimate leadership
One of the most important measures of leadership in Maharashtra resides in a concept called *shil*, which roughly translates as integrity. It is somewhat different from what we would call "character," mostly because the essence of shil is the correspondence between *saying* and *doing*, rather than the inherent moral correctness of either.[36] The difference here can be recognized in what one acquaintance once told me about the leadership of Mahatma Gandhi: "People weren't sure if he was good or bad, but they could see that what he said was what he did and *that* made him worthy of respect." Sharad Joshi, by being genuine and consistent in what he says and does, augments the public perception of his shil—and, by describing the government as an authority that rules through deception, he helps to diminish the already suspect shil of most conventional party leaders. This measure of leadership based on shil is important in its implication that a leader's legitimacy is not only based on public approval of his or her lifestyle, social roots, or declared agenda. To some extent, shil has the capacity to override these other considerations.

A second important expectation of a leader in Maharashtra is that he or she be a *mahapurush*, meaning a "great man" (or "great person"). A mahapurush does not have to be self-negating or a martyr (unless he *says* that is what he is)—he is measured for his ability to get things done, for the size and scope of those achievements, and for his ability to transcend petty rivalries that would prevent him from living up to his proclaimed intentions in accordance with his shil. Here again, Joshi is widely understood to meet the test. His perceived ability to make things happen—whether assembling mass audiences for a rally, enlisting participants in a specific agitation, garnering media attention, launching new initiatives, or negotiating concessions with members of the government—make him a mahapurush.

This status as a mahapurush and a man of shil partially explain why Joshi is able to command legitimacy as a rural leader and why he is widely respected with honorific terms such as *Bhao* (brother), *Rao* (the honorable) and *Saheb* (master or sir)—even among people who often disagree with him or who choose not to participate in the Sanghatana. Just as importantly, it also sheds some light on the easy association between Joshi and Bali. Bali, as a king who repeatedly defeated the gods and threatened their domain, is unquestionably a mahapurush. Moreover, in the trivikrama myth, Bali's shil is so unassailable that he gave up his whole kingdom in order to keep his word, even though he recognized that he had been deceived.

Another important cultural parameter of leadership that we must consider is the popular equation of a mahapurush with kingship and godliness. In Maharashtra, kings, leaders, elders, and everyday social superiors are often referred to in Marathi as "god"—and, in many social and discursive contexts, they are treated as such. In popular terms, this seems to extend from a conflation of both gods and social superiors as the ones who "provide the *roti*" (daily bread). In brahminical ideological terms, this ties in with the idea that social and spiritual hierarchy are effectively parallel—rendering one's social superiors always more godlike, or in closer proximity to god, than oneself. But even in popular (non-orthodox) representation, we see an almost fungible relationship between kings and gods, gods and kings. Just as the Varkaris' deity Vithoba is represented as the "king of Pandharpur," and Indra as "king" of the Puranic pantheon, many other gods are similarly represented as kings, and their images often bear scepters or crowns. Human kings, for their part, are often represented as gods, and their kingdoms as a microcosm of the cosmic hierarchy.

This relationship between gods and kings is not unique to Maharashtra (cf. Fuller 1992). One of the best examples of this is Lord Rama, who is understood by many people in India to have been a historical king, ruling from the city of Ayodhya in north India. In Maharashtra, we can also see an example of this in the deep veneration of Shivaji and his frequent representation alongside spiritual figures. The demon king Bali, though not a "god" by brahminical standards, also conforms to this model of both king *and* supernatural object of veneration. These same associations extend down through the social hierarchy, so that social superiors—a brahmin, a political leader, a boss, or a landlord—may frequently be addressed as a king or as a god. The key point to keep in mind here is that the public representation of Joshi as "god"—whether he is regarded as a human mahapurush who leads a community of subjects, or as an incarnation of king Bali—is not extraordinary in Maharashtrian culture. At the same time, however, such representations of ordinary human social superiors as deities are not necessarily to be taken literally. It is a cultural association that, depending on the beholder, may be as much a matter of convention as of a strategy aimed at flattering superiors in accordance with the cultural idiom.

There is one more factor we must consider in order to make initial sense of Joshi as a leader—not to mention one who is publicly venerated well above the status of most ordinary leaders. As an outsider to rural society and culture, Joshi represents no particular cultivating or laboring caste, no particular rural region and no particular sociopolitical loyalties. Unlike many other political leaders, Joshi has not cloaked himself in any of the conventional signifiers of specific caste or class struggles. Arguably, and contrary to his depiction as an unlikely rural leader, Joshi's strength as a rural leader and as a symbol of the movement is his public distance from any of those agrarian categories. Because he defies easy placement, he not only transcends rural social divisions and ideologies but is, in effect, an open sign that participants can attempt to endow with qualities and meanings that fit their own interpretation of who he is and their own expectations of him as a leader.[37]

All of this helps make sense of how an outsider like Sharad Joshi could be regarded and respected as a leader, but we still have not reached a rational explanation for the extent of Joshi's veneration. To do that, we need to consider the potential range of meanings for Joshi as a symbol alongside the range of meanings that are carried by the demon king Bali.

Venerating Joshi as Bali

Why would people actually put pictures of Sharad Joshi in places that are more commonly reserved for spiritual figures and recognized gods? How can we explain a degree of deification that goes so far beyond the ordinary respect or strategic kowtowing that is normally bestowed upon leaders? At one level, we can understand this in terms of the cultural model of placing superiors in the symbolic role of deities. But the extent of Joshi's public veneration by so many participants means we must seek additional explanation. I would suggest that equating Sharad Joshi with Bali serves a number of different objectives in the struggle among participants to define the Shetkari Sanghatana, its agenda, and its community.

One of those objectives is to symbolically "naturalize" Joshi as a rural insider. It achieves this, in part, by de-brahminizing and de-urbanizing Joshi—both of these are qualities not easily associated with an agrarian identity, and they are antithetical to representations of Bali. In the process, this also erodes the legitimacy of brahminism and urbanism as bases for any rural authority. A second objective is to endow Joshi with new bases of authority that are decidedly rural. Associating Joshi with Bali lifts the movement's leader up to the symbolic status of kingship and godliness (or at least a virtuous supernaturalness, depending on one's perspective) that characterizes the demon king himself. In these ways, associating Sharad Joshi with Bali Raja can be seen to benefit the movement at large—its leaders and its community of participants—by helping to establish a collective perception of Joshi as a legitimate *and rural* leader.

If we look more closely we also recognize a number of ways in which this ongoing association may be of particular significance to the lower-caste kunbi and Dalit masses of rural Maharashtra. It seems clear that advancing Sharad Joshi as Bali and promoting the use of Bali generally as a core symbol for the movement are directly opposed to the anti-Bali discourses and practices of some actors among the upper castes. Any insult to Bali, then, becomes an insult to the movement and its leader. The effect is one of putting the strength and reputation of the entire movement squarely behind the defense of Bali—and this is especially powerful when we recall that Bali is closely identified with the low-caste members of the rural community.

The association of Bali and Joshi also serves to further project Bali into the public sphere of rural life. This is especially important when we recall the ways that Bali is symbolic of opposition to Puranic authority and is metaphoric of a wide range of low-caste spiritual practices and supernatural understandings. Putting Joshi's image in the godhouse or anywhere else typically occupied by images of Puranic deities is equivalent in some respects to putting Bali in these locations—validating subaltern demon veneration alongside Puranic worldview.

Beyond all this is one further significance that is far more radical: equating Sharad Joshi with Bali is, for members of the lower castes who themselves identify with Bali, tantamount to equating the central leader of the movement with the low-caste self. In a sense, this can be seen as an extension of the way that shudra agriculturalists address one another as Bali, "the king of demons and shetkaris." It is also comparable in many respects to the special agrarian treatment of bulls and bullocks as Bali *and* as a king during the festival of Pola. It is important to recall here that the worship of the animal during Pola is also a sort of veneration of the self. Thus, placing an image of Joshi amongst the Puranic deities is comparable on a certain level to placing the subaltern self among the gods, to be venerated by the larger community of villagers—or the larger community of the Shetkari Sanghatana—including the rural elite. This conceptual equation between Joshi and the least advantaged members of the Sanghatana community can also be seen as a symbolic effort to "naturalize" the meaning of Joshi's rural insider status in line with a particular contour of rural identity—that of the most precariously positioned agriculturalists and laborers, who are predominantly members of the lower castes.

In this last sense, the discursive and practical enshrinement of Joshi as Bali can be seen as part of a larger struggle to build a beneficial alliance with a leader who demonstrates integrity and the capacity to make things happen. It is one more important expression of an overall strategy for projecting the interests of the relatively disenfranchised members of the rural community into the ideology and objectives of the movement.

Conclusions

In the previous chapter I described how participants in the Shetkari Sanghatana have substantial opportunity to influence the movement's objectives and the specific issues that it undertakes. In this chapter, I have argued that participants also actively influence the meaning and experience of Sanghatana culture and collective identity. As a window into this co-creative process, we have focused on the idiom of Bali Raja as the one major component of the Sanghatana's complex of signs and meanings that more than any other evokes the Sanghatana as a whole—not just the movement as an abstract phenomenon, but also its prescriptions and diagnoses, its spatiality, its participants, and its leaders.

The first part of the chapter considered some of the ways in which the myth of Bali's defeat holds distinctive importance in rural Maharashtrian culture. I then explored variant interpretations of the Bali myth as embodied in classical texts, oral folklore, rural speech and ritual practices, and considered their implications for differently positioned rural subjects. Next I examined ways in which the Bali myth evokes rural unity as well as rural differentiation and competition, and we saw that these same terrains of meaning and identity—terrains that are sometimes overlapping, sometimes conflicting—are as present within the Sanghatana community of participants as they are within rural Maharashtrian society in general. Finally, the chapter considered a number of ways in which the Sanghatana can be understood as a resource through which actors have managed to project some of the most subaltern and peripheralized of these meanings into the public sphere and into the active dialogue of the movement community. This is done through public cultural productions that lend emphasis to desired meanings, as well as through competitive manipulation of the meaning of the movement's paramount leader.

As we have seen, ground-level participants are actively engaged in the production of Bali's meaning within the movement. Moreover, their engagement with the Sanghatana idiom of Bali does not merely reproduce a movement meaning-complex that serves the interests of a dominant and empowered minority. On the contrary, participants' ongoing production and reproduction of this idiom represents real opportunities for, and often deliberate competition over, the construction of the idiom's meanings and the movement it signifies.

The Bali myth functions as an effective movement idiom for several reasons. First, it comprises a set of cultural signs that are widely perceived as deeply Maharashtrian and deeply rural. Second, it is capable of expressing and reflecting rural Maharashtrian experiences in ways that are not only effectively overlaid with the Sanghatana's widely articulated diagnoses of the rural condition and its prescriptions for change, but also in ways that—like these diagnoses and prescriptions—are applicable to a broad cross section of real subject experiences and amenable to variant interpretations or evolutions of the movement itself. Third, and most importantly, the social significance of Bali is itself open to a wide range of interpretations and, thus, the sign is capable of carrying widely variable meanings whenever it is invoked.

Contrary to notions that the most effective ideological signs are those that are the most hegemonically defined and univocal, we can see that the myth of Bali is effective as a movement idiom because of—not despite—its multivocality and ambiguity. This ambiguity gives it the potential to signify, within the movement idiom, not just the contours of consensus within the movement community, but also the borderlands of disagreement and discontent. In this way, the idiom embodies not only the complex collection of shared and divergent identities and interests that characterize the social composition of the movement collectivity—it also embodies a broad spectrum of subjects' interpretations of what the movement is, whose interests it does or could represent, and all that it has the potential to become.

Notes

[1] Bali does appear in numerous different classical myths. However, what I refer to here as "the" Bali myth is the story (and its variations) of Bali's deposition and his banishment to Patal. This is the best-known story of the demon king, and is the story in which Bali plays his most central role.

[2] The significance here is a matter of degree, and I do not mean to suggest that Bali is not invoked in cultural expressions in other areas of the country. Shulman (1980), for instance, indicates that Bali plays a recurring role in oral temple myths in Tamil Nadu, and Bali also holds a position of some significance in Orissa and parts of Andhra Pradesh. However, the role of Bali in these states is neither as publicly ritualized as it is in Kerala, nor as recurrent in daily village culture as in Maharashtra. It is interesting to note here that Kerala and Maharashtra are, to my knowledge, the only two Indian states in which the word "Patal" (the nether world to which Bali was banished) can also refer to distant regions at the outer edge of state's linguistic area. Just as Patal is understood by many Maharashtrians to mean "the Konkan," in Kerala it can likewise mean "Konyakumari"—the area at the southern tip of the state (see Joshi, Pandit Mahadevashastri 1962).

[3] Onam, widely celebrated in Kerala, is an annual festival commemorating events tied to Bali's downfall. According to Malayali accounts of the Bali myth, after Vamana wrested the kingdom from Bali the king persuaded Vishnu to let him return to visit his subjects once a year. Permission was granted, and this day of visitation is called Thirounam day—the final day of the ten-day festival (Verma 1997). In recent years, the spread and popularity of the Onam festival—much like the Ganapati festival in Maharashtra—has become entangled with state government efforts to publicize and support public expressions of essentialized Malayali culture, in part for the promotion of tourism.

[4] *Gavagada*, Atre's Marathi account published in 1915, is a pioneering study of the *balutedari* system of hereditary village service in rural Maharashtra. My appreciation goes to Sumit Guha and Pramodh Kale for directing me to this text.

[5] I should be clear here about what I mean by hierarchy. In talking about "upward" and "downward" shifts in expressed identity I do not mean to suggest that hierarchy is a "natural" and un-interrogated structure of mind in the sense articulated by French anthropologist Louis Dumont (1970). Rather, I view this hierarchy as a historically constituted and

continually contested structure of relative power that is articulated and experienced through symbolic constructs as well as through differential access to social and material resources. Shifts in expressive identity are just one way that social hierarchy is accommodated, negotiated, or contested by individuals and groups.

[6] *Shetmazur* means "agricultural laborer." The movement, which was active in the districts of Sangli, Satara, Solapur, and Kolhapur at the time of my research, had no constructive relationship with the Shetkari Sanghatana (even though there is some overlap in their areas of organizing). According to one activist in the Shetmazur Shetkari Sanghatana: "We are not ideologically opposed to Joshi's Shetkari Sanghatana, but we see our own work as being more focused on labor."

[7] I have adapted the following from the main elements of the myth as given in Mani's *Puranic Encyclopedia* (1975). The specific verses on which this adaptation is based are found in the Vamana Purana, [75–76, 77] and the Bhagavata [Skandha 8].

[8] "Trivikrama" refers both to the story itself and to the giant, three-stepping avatar of Vishnu assumed as he morphed from the dwarfen form of Vamana. The story has both Vedic and Puranic origins, as does Bali, but the trivikrama as a myth of Bali's downfall is distinctively Puranic.

[9] The common Maharashtrian name for the elephant-headed deity known as the "creator and remover of obstacles." Ganapati (known and pronounced more generally in other parts of India as *Ganesha*) is much revered in Maharashtra.

[10] Encyclopedic works drawing on original Puranic sources generally describe the first step as "the entire earth" and the second as "the universe." All apparently agree that the only place remaining for the third step was Bali's head, but the ancient accounts differ on whether or not Vamana actually took the third step. In at least one classical account, Vishnu/Vamana, in return for Bali's performance of his *dharma* by making sacrifices to both god and brahmin, chose not to take the third step plunging Bali into Patal—but rather granted Patal to Bali as a kingdom in exile and installed him as its monarch (Cf. Mani, Vettam 1975, *Puranic Encyclopedia*; Joshi, Pandit Mahadevashastri 1962, *Bharatiya Sanskrutikosh*; Chitrao, M. M. Sidheshwar Shastri 1964, *Bharatavarshya Pracin Caritrakosh*).

[11] A similar argument has been made with regard to other Hindu myths, very notably the stories embodied in the Ramayana epic. See, for example, contributions to Richman (1991; 2000).

[12] As with other major cultural signs that we have considered in the Sanghatana constellation, it also signifies *differentiation* (distinguishing the Sanghatana from movements and policies that benefit India at the expense of Bharat) and *continuity* (allying the Sanghatana with an ongoing struggle for rural autonomy and equality that stretches back to mythical history).

[13] In one such tale that I heard in a village in Marathwada, the striking resemblance of the story to the trivikrama myth also illustrates the agricultural castes' self-identity with Bali Raja and his fate. One day, according to the story, Lord Shiva and his consort Parvati were having an argument about the dutifulness of the shetkaris. (My informant could not recall anything more about the argument, or the differing opinions of the two deities.) To settle the matter, Shiva disguised himself as a beggar, wandered into a cultivator's field, and stole some of the crop for food. The cultivator ran out, scolded and beat the beggar. Suddenly, the beggar transformed himself back into Shiva, and Parvati also appeared on the scene. Parvati said to the cultivator: "Why do you behave like this with a poor beggar? You are supposed to be Bali Raja, generous and kind!" The cultivator realized his error, and begged *ushappa* (mitigation, mercy) from Parvati. She relented, and punished the cultivator with a relatively mild curse: that from now on, the four corners of any villager's field will not be equally productive—one corner will always have a poor yield. According to my informant, the curse was intended to remind the shetkaris to always be generous.

[14] The Vedas, the earliest of the classical Hindu scriptures, are collections of Sanskrit hymns and ritual procedures comprising four volumes. The oldest of these, the Rg Veda, is thought to have been composed in its currently known form somewhere between 1500 and 1000 BC. The Puranas, comprising tangled and multilayered stories about the gods, in a collection of eighteen major texts and a disputed number of minor texts, date to a period between 300 and 1000 AD.

[15] In the Puranic literature, Lord Indra—the god of the Rg Veda who brought nourishing rains and was the benefactor of the pastoral Aryans and their cattle—is sidelined. Although Indra retains a legitimizing role as the king of all gods and as the ultimate ruler of the heavens, he (as with the other Vedic gods Agni, Soma, Vayu, and Surya) is no longer

represented as a god for the *people*. Rather, he is represented as a (somewhat ineffectual) chief of the newer gods—the Puranic deities who have become the gods with whom people more commonly interact. This demotion is clearly evident: Puranic myths often depict Indra as a figure who cannot act decisively to retain his domain without the assistance of the new pantheon; hold him accountable for the murder of a brahmin; and often describe how he is lambasted by gods and brahmins alike for his arrogance (Dimmitt and van Buitenen 1978).

[16] A somewhat analogous example of this shift can be found in the emergent monotheistic conception of god among the ancient Hebrews. The new god of the Hebrews amalgamated many deep-rooted cultural understandings of divinity, but was also intolerant of other pre-existing forms and practices such as the worship of golden calves and graven images that were not in his likeness.

[17] The Sanskrit term asura applies to a range of demonic beings in the Maharashtrian supernatural world, including beings known as *daityas, rakis,* and *rakshasas*. Other beings that are often popularly recognized as having an affinity with asuras are ghosts and goblins (*bhuts, pretas*).

[18] This transformation must be understood in the context of Hindu theodicy, and cannot be simply taken as a parallel to the notions of "good" and "evil" in some other world religions. Even in the Puranic literature, demons are rarely depicted as thoroughly and essentially evil, and gods are not thoroughly and essentially good (see, for example, O'Flaherty 1976). Equally challenging are the facts that the killing of a demon is sometimes depicted as sinful, even if necessary, and that the gods themselves sometimes express identity with the demons they kill or oppose. Thus, the degree to which demons are represented as evil or a threat to gods and humans is a relative one rather than a fully polarized opposition.

[19] Translated from the Vamana Purana [47:1–51; 48:1–15].

[20] Translated from the Matsya Purana [128–139].

[21] One need only consider the 1992 destruction of the Babri Mosque in Ayodhya that was informed by mythic accounts of the life of Lord Rama, for example, to recognize the role that myth can play in structuring "truth" and orienting contemporary models for action.

[22] We can say this not only with regard to Hinduism proper, but also, to some extent, to other religious systems and communities that have been in close contact with the Hindu world. Consider, for example, the

existence of caste as a functioning social rubric within Maharashtrian and other Indian Muslim and Christian communities.

[23] For this reason, my use of the word "orthodox" anywhere in this writing will always refer only to *claims* of orthodoxy by situated practitioners.

[24] Lingayats are members of a Hindu sectarian community founded in Karnataka in the twelfth century. Originally comprising members of shudra castes that joined the sect, in Maharashtra the community today functions as a corporate caste group (with recognized subcastes within the community) and membership is ascribed by birth. Though originally shudra, Lingayat men and women customarily wear the sacred thread—a symbol of identity more commonly associated with the brahmin community. As "Vivek" described his community: "We are just like brahmins."

[25] Folk goddesses are commonly associated with diseases.

[26] Religion scholar Anne Feldhaus (1995), for example, has noted that many of the classical Puranic tales narrating the killing or defeat of demons are commonly drawn into localized Marathi textual and village oral narratives about the origins of sacred geographies in Maharashtra. Battles with demons, thus, are seen as a precursor to establishing the sanctity of a landscape in which people live and with which they interact. In her work, Feldhaus cites an extensive array of battle stories that are drawn into these narratives by agriculturalists, including—but not limited to—Vishnu's defeat of Bali. Other popularly recounted tales of conquest include Vishnu's defeat of the demons Karaka, Madhu and Kaitabha; Vishnu's or Shiva's defeat of Andhaka; Shiva's killing of the three Tripurasusas; Lord Ganapati's killing of Gajasura; Rama's killing of the giant demon Ravana from the land of Lanka; Bhavani's bloody defeat of the buffalo demon Mahishasura who threatened the stability of the world; Narasimha's killing of the demon king Hiranyakashipu for his renunciation of Vishnu; Lord Krishna's killing of Vajraketu; and Indra's killing of the demons Vrta, Namuci, Bala, and Kadamba. This list only touches the surface of an extensive body of popular Marathi written and folkloric accounts that correspond with classical narratives about securing the safety of the world and eliminating evil through the systematic killing or banishment of demon beings.

[27] Durga is a powerful warrior deity, specifically created by the other gods for the purpose of killing demons.

[28] It is interesting to note that most major commemorations of demon conquest occur in the same short period of the annual cycle, the tail end of a particularly inauspicious period in the Hindu ritual calendar during which Vishnu is said to have been sleeping for the previous four months. In myth as well as in actual festival practice, the divine context is one in which Vishnu has awakened to discover that demonic forces have gathered their strength and extended their power over the brahminical world during his slumber. The festivals thus, are a reassertion of order—not only the divine order that Vishnu is charged with maintaining, but also the village social order of rulers and differentially ranked subjects, that is a microcosm of the cosmic hierarchy (Fuller 1992).

[29] The actual duration depends upon the lunar circumstances in a given year.

[30] Maruti is the popular Marathi name for Lord Hanuman, the "monkey god" who is a devotee and companion of Lord Rama. Nearly every Maharashtrian village has a Maruti temple.

[31] Although *beil* literally means "bull" or "bullock," many low-caste laborers or pastoralists who do not own draught animals commonly celebrate Pola as well, venerating whichever animal performs services for them. Many noncultivators venerate donkeys on Pola—for example, potters, diggers, or stonebreakers who depend on donkeys for hauling. Some people also venerate dogs. As one informant explained, dogs perform services for the very poor by protecting the community perimeter and barking at intruders. Dogs are also symbolic of Khandoba (who has a dog as his hunting companion), and this may play a role in the extension of Pola to dogs.

[32] This rising interest in the public invocation of Bali in Maharashtra may fit into what seems to be a larger national trend of public demon worship—possibly revealing a larger pattern of deliberate re-complications of the spiritual landscape in the public sphere. Among lower castes and women's groups in India there has been a growing public dialogue on the virtues of many of the demons killed off in Puranic myth. One group for example—the Ravan Darshan-Igdarshan Manch, in Uttar Pradesh state—has planned to construct India's first known temple to the demon Ravana. The head of the Manch, quoted in the *Times of India*, said that the temple would serve as a sort of "demon culture" center, dedicated to countering "the forces of casteism, corruption and communalism" and replacing them with "the ancient concept of

nationalism which unites the Vedic and non-Vedic schools of culture in India" (TOI, 5/8/98, p. 2).

[33] In the Ramayana, Rama subjects his wife Sita to a test of purity in order to determine if she had been unfaithful (whether willfully or by force) while in the demon Ravana's captivity. In the epic story, Rama appears unwilling to accept Sita's own account of her faithfulness, and is prepared to reject her even if she had been unfaithful only by force.

[34] The contrast with many other Indian rural leaders here is striking. Mahendar Singh Tikait of the BKU movement in Uttar Pradesh maintains a very strong cultural, political, and sartorial connection to his family clan and his Jat caste—and is generally regarded as a Jat leader as much as an agrarian leader. Professor Nanjundaswamy, the leader of the KRRS agrarian movement in Karnataka is a devout wearer of *khadi* (handwoven cotton) clothing. This is an overt symbol of the Gandhian ideal of supporting rural products and village-based cottage industries (as opposed to urban-industrial manufactured artificial fibers). Even Mahatma Gandhi—who initially returned to India in the suit and tie of an urban, educated sophisticate—went through several iterations of costume changes before finally donning the simple loincloth that he felt best evoked his sympathies for the rural masses. (For a fascinating discussion of Gandhi's clothes see anthropologist Emma Tarlo 1996.)

[35] This contrasts significantly with most literature on millenarian and revitalization movements, a category that otherwise has some interesting parallels with the Sanghatana. Adas (1979), for example, describes the leaders of the millenarian movements in his comparative study as people who believe that they have special contacts with supernatural forces, and are able to convince others of this. Joshi has no such pretenses and, to the extent that he is identified as a man with special supernatural or divine characteristics, this perception and assertion is not Joshi's own. This suggests a potentially rich area for further research on millenarian leaders: who decides that the leader has powers or characteristics that exceed those of ordinary mortals?

[36] Thanks to "K.S." who introduced me to this difference.

[37] Dipankar Gupta (1997) makes a similar point, suggesting that agrarian leaders who come from outside rural areas are free from acquired social baggage.

6—Conclusion: Cultivating community

The Shetkari Sanghatana brings attention to a number of important implications for how we think about collective cultural and political expressions, particularly in the context of a broad-based social movement. In this brief conclusion, I will summarize some of these implications.

Participation and participants

In chapter 1, I introduced a dominant theme in the literature on social movements that I call the *premise of progressive solidification*. This premise suggests that as a social movement draws greater numbers of participants it is characterized by increasing consensus around its ideology and collective objectives. The case of the Shetkari Sanghatana suggests that the premise of progressive solidification overextends the primacy of consensus in a way that inhibits our understanding of a mass movement's plurality of subjectivities and the nature of subjects' agency.

One of the core arguments of this book has been that successful mass social organizing may be characterized at least as much by contextually shifting participant agendas and fluctuating participation as by a unity of ideas or a tightly defined participant base. The Shetkari Sanghatana is, in most respects, socially and geographically amorphous and subject to wide variances of interpretation by its participants. Participants shift in and out of the movement, engaging with it to a greater or lesser degree in line with their own assessments of the advantages and disadvantages of participation. At any given time, in any context, participants in the Sanghatana community may experience some degree of consensus on the movement's objectives—but at other times, in other contexts, they may find themselves to be substantially in disagreement. As I have argued, participatory shifts and disagreement are not failures or hindrances for the movement: they are characteristics that enable its continued social expansion and sustainability.

Part of the reason we often fail to recognize these fluctuating terms of consensus and opposition within broad social movements is that we are sometimes insensitive to the ways in which the contours of identity and social participation within society at large are contextually shifting and strategically motivated. For this reason, throughout this book I have sought to emphasize these shifting contours in relation to a number of different, often reified, categories—including geography, religion, caste, and class.

In the context of a mass rural social movement, class in particular is a category that has served to uphold the progressive solidification paradigm. One major expression of this has been the theoretical and methodological differentiation between "peasants" and "farmers" as distinct classes. There have been great debates among scholars on the specific attributes that characterize one over the other. In the literature, the distinction between peasants and farmers tends to lie either in different value systems and cultural lives or in different relations of production. While these differences are useful in objective analysis, they may mislead our understanding of people's actual subjective experiences. As I examined in chapters 2 and 3, rural Maharashtrians do, in many contexts of social and political engagement, actively resist the peasant vs. farmer distinction. Just as importantly, they often strategically align themselves and their performed identity with a socioeconomic status that may not correspond with their objective class.

These contextual shifts and fluctuations in the performance and experiences of socioeconomic identity make it important for us to understand both the *appearance* of collective unity and the *reality* of differentiation in ways that are attentive to subjects' own understandings. It is their own understandings of the world and the range of opportunities it affords that inform subjects' choices for social and political action.

Movement narratives and ideological frames

This kind of ethnographic approach to participants' own experience and motivation for action suggests that we look more closely at the ways we analyze collective narratives and ideologies. As I have suggested, the signs and idioms used within the Shetkari Sanghatana are effective, in part, because of their substantial ambiguity of meaning and their multivocality. Shetkari Sanghatana participants are highly aware of the interpretive variability that characterizes not only their own action but also the core symbols associated with the movement. Moreover, participants are, to a large extent, actually empowered by this ambiguity, which creates opportunities for overt and covert dialogue on the meaning of the movement and for participatory crafting of the movement. Contrary to the solidification paradigm, what we see in the Sanghatana is a confluence of broad ranging actors who do not fully agree on the exact nature of their interests or the precise boundaries and meaning of their solidarity. In the Sanghatana, these actors have coalesced into a pattern of engagement that is more or less stable but also continuously evolving and dialogical.

A number of anthropologists have shown us ways that cultural narratives and symbols of collective identity derive their strength from ambiguity. Anthropologist Kirin Narayan (1989), for example, in examining the role of stories as a medium of religious instruction, finds that much of what makes a guru's storytelling effective is the opportunity it affords for each listener to interpret the story in a way that best fits the listener's own personal situation and his or her broader understandings of the world. These interpreters may consider themselves members of a larger community of interpreters (for example, a specific community of Hindu religious pupils, or even a more generalized community of spiritual seekers), but their interpretations are largely individual and personal, even as they are also, on another plane, substantially aligned with many meanings that may be common to the broader group. In a more political context of collective meaning, we can turn to anthropologist Eric Wolf's

(1958) analysis of the Virgin of Guadalupe as a symbol of Mexican national identity. Wolf finds the image of Guadalupe's Virgin a "Mexican master symbol" (34) because of its capacity to separately but simultaneously reflect the often-opposed identities of ethnically Spanish Mexicans *and* indigenous Mexicans. Wolf goes on to define such master symbols as those that "provide the cultural idiom of behavior and ideal representations through which different groups of the same society can pursue and manipulate their different fates within a coordinated framework" (ibid).

In any mass collective embrace of what Wolf calls a master symbol, the interpretations of individuals, representing differently positioned social groups, are sometimes competitive in relation to other interpretations—but they can also be accommodative of the interpretation of the larger collectivity. In the case of this Mexican sign, the power to signify and validate differential *and* national identities is what makes the sign effective for the larger collectivity but also empowering for each group within it. In Maharashtra, Shivaji and the Varkari Sampradaya can be similarly understood as such master symbols.

Within the Sanghatana community, the master symbols are Bali Raja and Sharad Joshi. Just as in the case of a multi-ethnic nationalist identity, the fact of concerted mass involvement in Sanghatana rallies and agitations should not be taken as proof that participating subjects are in complete agreement on their vision of the world, or that they have subsumed other understandings of their individual or collective selves within the mass collective selfhood of the movement. This same cautionary rule about deducing supposed realities of agreement from observed performances of solidarity is one that may apply equally to movements and collectivities that are much smaller than the Shetkari Sanghatana. As anthropologist James Fernandez (1965) has demonstrated in his classic analysis of the Bwiti spiritual cult among the Fang people of northern Gabon, collective performances of consensus may contribute to both the real and the apparent solidarity of a group *despite* the fact that participants may share only a weak consensus on the actual meaning of their specific performative acts.

Leaders and leadership

These insights bring us to a final set of implications with regard to how we think about the leaders of mass social movements. All of the above suggests that movement leaders are significantly something less than hegemonic dictators of movement ideas and objectives, and yet we also know that leaders are something more than regular participants.

There is an emerging body of literature supporting a perspective in which leaders are not so much the apical sources of a movement's ideas or the principle founts of its participants' inspiration, but are resources that are partially defined and partially manipulated by the community of people who "follow" them. Leadership theorists Sonia Ospina and Ellen Schall (2001) argue that leaders are often better understood as emergent social products rather than simply as a set of attributes and behaviors that are strictly determined by leaders themselves. Because my research on the Sanghatana led me to a similar conclusion, I suggest that we mostly avoid terms such as "follower" or "member" when talking about a movement's grassroots base, opting instead for "participant." The term participant leaves open the possibility of greater co-production by actors engaged in the movement. The term participant is also more true to the fluctuating nature of most actors' real involvement—shifting in and out of formal and informal roles, and greater or lesser engagement. This idea that both movements and leaders are co-produced can also be found in the work of sociologist Belinda Robnett (1997). Robnett uses the term "bridge leader" to describe movement participants who serve as a link between central leaders and the various constituencies of a movement's base. In her description, bridge leaders, help orient more formal leaders and shape the culture and organization of the movement from the ground up. Comparable roles in the Shetkari Sanghatana are those I have referred to as activists, local organizers, and district-level leaders. Actors in these roles are important conduits for information and ideas between central leaders and grassroots participants, in both directions.

Despite the collaborative, dialogical construction of objectives and meaning in the Shetkari Sanghatana, a surface view of the movement can easily give the impression of a clear boundary between leaders and followers—especially in the case of Sharad Joshi. One reason for this is that participants frequently speak of Joshi as a divine or semi-divine being. Another reason is the Sanghatana's idiom of restoring of a mythical golden age under king Bali. This gives the Sanghatana overtones of a millenarian or revitalization movement—types of movements that are generally described to have a visionary, charismatic leader at the head of loyal, unquestioning followers (cf. Adas 1979; Peires 1989; Wallace 1956). The Sanghatana casts a fresh light on such movements in at least two significant ways. First, as we have seen, the Sanghatana's millenarian vision is, for the vast majority of participants, a *metaphor* for change rather than an actual hoped for or expected outcome of the movement's success. Second, participants' characterizations and interpretations of Joshi vary substantially. Even when Sharad Joshi is projected as a divine or semi-divine being, this most often reflects a cultural association between leadership and divinity rather than a genuine perception that the leader is supernatural. It also reflects strategic efforts to localize the leader, build reciprocal ties with the leader, or align the leader with specific contours of selfhood.

Thus, the Sanghatana case suggests that it may be important to rethink our understanding of leaders—even those we understand to be impressively charismatic and revered. Leaders are, significantly, products of the populations they lead. When these populations are socially diverse, social competition in the creation of leaders may play a very important role in determining the specific attributes through which leaders are signified. In the final analysis we may find that leaders not only have less control over the cultivation of movement agendas, movement cultures, and movement communities than we have theorized in the past—

they may also have less control over the cultivation of their own role and meaning as a leader.

Cited and consulted works

Cited periodicals
Deshonatti (in Marathi)
Indian Express (IE)
Maharashtra Times (MT)
Sakal (in Marathi)
Shetkari Sanghatak (in Marathi)
The Times of India (TOI)

General bibliography
Abu–Lughod, Lila. 1990. "The Romance of Resistance: Tracing Transformations of Power Through Bedouin Women." *American Ethnologist* 17, no. 1, 41–55.

Adas, Michael. 1979. *Prophets of Rebellion: Millenarian Protest against the European Colonial Order*. Chapel Hill: University of North Carolina Press.

———. 1981. "From Avoidance to Confrontation: Peasant Protest in Precolonial and Colonial Southeast Asia." *Comparative Studies in Society and History* 23, no. 2, 217–47.

———. 1991. "South Asian Resistance in Comparative Perspective." In *Contesting Power: Resistance and Everyday Social Relations in South Asia*, edited by Gyan Prakash and Douglas Haynes, 290–305. New Delhi: Oxford University Press.

Alvarez, Claude. 1992. *Science, Development and Violence: The Revolt Against Modernity*. New Delhi: Oxford University Press.

Ambedkar, Babasaheb. 1979. "On Linguistic States." Vol. 1, part 2 of *Writings and Speeches*. Bombay: Department of Education, Government of Maharashtra.

Anderson, Leslie E. 1994. *The Political Ecology of the Modern Peasant: Calculation and Community*. Baltimore, MD: Johns Hopkins University Press.

Appadurai, Arjun. 1984. "How Moral is South Asia's Economy?" *Journal of Asian Studies* 43, no. 3, 481–97.

———. 1990. "Technology and the Reproduction of Values in Western India." In *Dominating Knowledge: Development, Culture and Resistance*, edited by Frederique Apffel Marglin and Stephen A. Marglin, 185–216. Oxford, UK: Clarendon Press.

———. 1991. "Global Ethnoscapes." In *Recapturing Anthropology*, edited by Richard G. Fox, 191–210: Santa Fe, NM: School of American Research Press.

Athreya, Venkatesh, Goran Djurfeldt, and Staffan Lindberg. 1990. *Broken Barriers: Production Relations and Agrarian Change in Tamil Nadu*. New Delhi: Sage.

Atre, Trimbak Narayan. 1915. *Gavagada*. Pune: Aryabhushan Press (in Marathi).

Attwood, Donald W. 1979. "Why Some of the Poor Get Richer: Economic Change and Mobility in Rural Western India." *Current Anthropology* 20, 495–516.

———. 1988. "Risk, Mobility and Cooperation in Maharashtrian Villages." In *City, Countryside and Society in Maharashtra*, edited by D. W. Attwood, M. Israel, and N. K. Wagle, 172–90. Toronto: Center for South Asian Studies.

———. 1992. *Raising Cane: The Political Economy of Sugar in Western India*. Boulder, CO: Westview Press.

Babar, Sorojni: 1968. *Folk Literature of Maharashtra*. New Delhi: Maharashtra Information Center, Government of Maharashtra.

Badgaiyan, S. D. 1993. *Industrialization and Peasant Social Formation*. Delhi: Rawat.

Banaji, Jairus. 1977. "Capitalist Domination and the Small Peasantry: Deccan Districts in the Late Nineteenth Century." *Economic and Political Weekly* 12, no. 33–4 (August): 1375–1404.

Bardhan, Pranab. 1984. *The Political Economy of Development in India*. Oxford: Blackwell.

Basso, Keith H. 1979. Portraits of "the Whiteman". Cambridge, UK: Cambridge University Press.

Basu, Amrita. 1986. "Contrasting Modes of Agrarian Protest in India: The Significance of Gender, Ethnicity and Class." PhD diss., Columbia University.

Bateson, Gregory. [1954] 1972. *Steps to an Ecology of the Mind*. New York: Ballantine.

Benford, Robert D. 1993. "Frame Disputes within the Nuclear Disarmament Movement." *Social Forces* 71, 677–701.

Benford, Robert, and David A. Snow. 2000. "Framing Processes and Social Movements: An Overview and Assessment." *Annual Review of Sociology* 26, 611–39.

Beteille, Andre. 1965. *Caste, Class and Power: Changing Patterns of Stratification in a Tanjore Village*. Berkeley: University of California Press.

———. 1991. "Introduction." In *Society and Politics in India: Essays in a Comparative Perspective*, edited by Andre Beteille, 1–14. London: Athlone Press.

Bhaba, Homi K. 1990. "DissemiNation: Time, Narrative, and the Margins of the Modern Nation." In *Nation and Narration*, edited by Homi K. Bhaba, 291–322. London: Routledge.

Bhagwat, Durga. 1958. *An Outline of Indian Folklore*. Bombay: Popular Book Depot.

———. 1974. "The Vithoba of Pandhari." Translated by Gunther D. Sontheimer. *South Asian Digest of Regional Writing* 3, 112–20.

Bloch, Maurice. 1986. *From Blessing to Violence: History and Ideology in the Circumcision Ritual of the Merina of Madagascar*. Cambridge, UK: Cambridge University Press.

Bourdieu, Pierre. 1977. *Outline of a Theory of Practice*. Translated by Richard Nice. Cambridge, UK: Cambridge University Press.

———. 1990. *The Logic of Practice*. Translated by Richard Nice. Stanford, CA: Stanford University Press.

Bouton, Marshall. 1985. *Agrarian Radicalism in South India*. Princeton, NJ: Princeton University Press.

Brahme, Sulabha, and Ashok Upadhyaya. 1979. "A Critical Analysis of the Social Formation and Peasant Resistance in Maharashtra," 3 vols. Unpublished manuscript.

Brahme, Sulabha, and R. P. Nene. 1985. *Maharashtratul Shetmajur* (Farm Laborers in Maharashtra). Pune: Phule Samata Pratishan (in Marathi).

Brass, Paul. 1991. "Moral Economists, Subalterns, New Social Movements, and the (Re-) Emergence of a (Post-) Modernized (Middle) Peasant." *Journal of Peasant Studies* 18, no. 2 (January): 173–205.

———. 1994a. "The Politics of Gender, Nature and Nation in the Discourse of the New Farmers' Movements." *Journal of Peasant Studies* 21, no. 3–4 (April–July): 27–71.

———. 1994b. "Introduction: The New Farmers' Movements in India." *Journal of Peasant Studies* 21, no. 3–4 (April–July): 3–26.

———. 2000. *Peasants, Populism, and Postmodernism: The Return of the Agrarian Myth*. London: Frank Cass Publishers.

Bremen, Jan. 1994. *Beyond Patronage and Exploitation: Changing Agrarian Relations in South Gujarat*. Delhi: Oxford India.

Brim, John A., and David H. Spain. 1984 [1974]. *Research Design in Anthropology: Paradigms and Pragmatics in the Testing of Hypotheses*. New York: Irvington Publishers.

Byres, T.J. 1994. "'Preface' to Special Issue on New Farmers' Movements." *Journal of Peasant Studies* 21, no. 3–4 (April–July): 1–2.

Calman, Leslie Joan. 1984. "Authority's Response to Challenge: Protest in Democratic India." PhD diss., Columbia University.

Carter, Anthony T. 1974. *Elite Politics in Rural India: Political Stratification and Political Alliances in Western Maharashtra*. London: Cambridge University Press.

———. 1988. "Land Transactions and Household Dynamics in Maharashtra." In *City, Countryside and Society in Maharashtra*, edited by D. W. Attwood, M. Israel, and N. K. Wagle, 151–71. Toronto: Center for South Asian Studies.

Chambers, Robert. 1983. *Rural Development: Putting the Last First*. Harlow, UK: Longman.

Charlesworth, Neil. 1972. "The Myth of the Deccan Riots of 1875." *Modern Asian Studies* 6, no. 4, 401–21.

Chatterjee, Partha. 1987. "The Constitution of Indian Nationalist Discourse." In *Political Discourse: Explorations in Indian and Western Political Thought*, edited by Bhikhu Parekh and Thomas Pantham, 249–65. New Delhi: Sage.

Chayanov, A. V. 1986. *The Theory of Peasant Economy*. Madison: University of Wisconsin Press.

Chishti, Sumitra. 1991. "Systemic Changes in the World—Hegemony and the World Economic Order in the Nineties." *Indian Journal of Social Science* 4, no. 1, 45–57.

Chitrao, M. M., Sidheshwar Shastri. 1964. *Bharatavarshya Pracin Caritrakosh*. Poona: Bharatiya Charitrakosha Mandal (in Marathi).

Chopra, P. N. 1998. *Religions and Communities of India*. New Delhi: Vision Books.

Cohen, Jean L. 1982. *Class and Civil Society: The Limits of Marxian Critical Theory*. Amherst: University of Massachusetts Press.

———. 1983. "Rethinking Social Movements." *Berkeley Journal of Sociology* 27, 99–113.

———. 1985. "Strategy or Identity: New Theoretical Paradigms and Contemporary Social Movements." *Social Research* 52, no. 4, 663–716.

Cohn, Bernard S. 1987. *An Anthropologist Among Historians and Other Essays*. New Delhi: Oxford University Press

Comaroff, Jean, and John Comaroff. 1991. *Of Revelation and Revolution: Christianity, Colonialism and Consciousness in South Africa*. Chicago: University of Chicago Press.

Coronil, Fernando. 1992. "Beyond Occidentalism: Towards Post-Imperial Geohistorical Categories." Unpublished manuscript.

Dandekar, Hemalata. 1999. *Men to Bombay, Women at Home: Urban Influences on Sugao Village, Deccan Maharashtra, India, 1942–1982*. Ann Arbor, MI: Michigan Papers on South and Southeast Asia.

Das, Veena. 1976. "The Uses of Liminality: Society and Cosmos in Hinduism." *Contributions to Indian Sociology* (n.s.) 10, no. 2, 245–63.

———. 1995. *Critical Events: An Anthropological Perspective on Contemporary India*. New Delhi: Oxford University Press.

Deleury, G. A. 1960. *The Cult of Vithoba*. Pune, India: Deccan College, Postgraduate and Research Institute.

Deshmukh, Dinkar. 1985. "Peasant Movement in Maharashtra, with Special Reference to Shetkari Sanghatana (1980–1985)." MPhil thesis, University of Poona.

Desrochers, John. 1991. "The Role of Social Movements." In *Social Movements: Toward a Perspective*, edited by John Desrochers, Bastian Wielenga, and Vibhuti Patel, 5–73. Bangalore: Center for Social Action.

Dhanagare, D. N. 1990. "Shetkari Sanghatana: Farmers' Movement in Maharashtra." *Social Action* 40, no. 4, 347–69.

———. 1994. "The Class Character and Politics of the Farmers' Movement in Maharastra during the 1980s." *Journal of Peasant Studies* 21, no. 3–4 (April–July): 72–94.

Dikshit, K. R. 1986. *Maharashtra in Maps*. Bombay: Maharashtra State Board for Literature and Culture.

Dimmit, Cornelia, and J. A. B. van Buitenen. 1978. *Classical Indian Mythology: A Reader in the Sanskrit Puranas*. Philadelphia: Temple University Press.

Dirks, Nicholas B. 1992a. "Introduction." In *Colonialism and Culture*, edited by Nicholas B. Dirks, 1–26. Ann Arbor: University of Michigan Press.

———. 1992b. "Castes of Mind." *Representations* 37 (Winter): 56–78.

Dumont, Louis. 1970. *Homo Hierarchicus*. Chicago: University of Chicago Press.

Dwyer, Kevin. 1982. *Moroccan Dialogues*. Baltimore, MD: Johns Hopkins University Press.

Eagleton, Terry. 1990. "Nationalism: Irony and Commitment." In *Nationalism, Colonialism, and Literature*, edited byTerry Eagleton, Fredric Jameson, and Edward W. Said, 23–42. Minneapolis: University of Minnesota Press.

Efremova, Irina. 1997. "Caste Profile of Maharashtra: Old and New." Paper presented at the 7th International Maharashtra Studies Conference, January 2–6, Pune, India.

Eliade, Mircea. 1959. *The Sacred and Profane: The Nature of Religion*. New York: Harcourt Brace Jovanovich.

Engblom, Philip C. 1987. 1987. "Introduction." In *Palkhi: An Indian Pilgrimmage*, by D. B. Mokashi, translated by Philip C. Engblom and Eleanor Zelliot, 1–30. Albany: State University of New York Press.

Esteva, Gustavo. 1987. "Regenerating People's Space" *Alternatives* 12, 125–52.

Etzioni, Amitai. 1964. *Modern Organization*. Englewood Cliffs, NJ: Prentice-Hall.

Feldhaus, Anne. 1988. "The Orthodoxy of the Mahanubhavs." In *The Experience of Hinduism: Essays on Religion in Maharashtra*, edited by Eleanor Zelliot and Maxine Berntsen, 264–79. Albany: State University of New York Press.

———. 1995. *Water and Womanhood: Religious Meanings of Rivers in Maharashtra*. New York: Oxford University Press.

Fernandez, James W. 1965. "Symbolic Consensus in a Fang Reformative Cult." *American Anthropologist* 67, no. 4 (August): 902–29.

Fish, Stanley. 1980. *Is There a Text in This Class?* Cambridge, MA: Harvard University Press.

Foucault, Michel. 1980. *Power/Knowledge.* Edited by Colin Gordon. Brighton, UK: Harvester Press.

Fox, Richard G, ed. 1990. *Nationalist Ideologies and the Production of Nationalist Cultures.* Washington, DC: American Anthropological Association.

Frank, Andre Gunder, and Marta Fuentes. 1987. "Nine Theses on Social Movements." *Economic and Political Weekly* (May 21) 1503–10.

Friedland, William H., Lawrence Busch, Frederick H. Buttel, and Alan Rudy, eds. 1991. *Towards a New Political Economy of Agriculture.* Boulder, CO: Westview Press.

Friedman, Jonathan. 1994. *Cultural Identity and Global Process.* London: Sage.

Friedrich, Paul. 1977 [1970]. *Agrarian Revolt in a Mexican Village.* Chicago: University of Chicago Press.

Fuller, C. J. 1992. *The Camphor Flame: Popular Hinduism and Society in India.* Princeton, NJ: Princeton University Press.

Gala, Chetana. 1997. "Empowering Women in Villages: All-Women Village Councils in Maharashtra." *Bulletin of Concerned Asian Scholars* 29, no. 2, 31–45.

Ghosh, Arun. 1990. *Agrarian Structure and Peasant Movements in Colonial and Post-Independence India.* Calcutta: K. P. Bagchi and Co.

Goffman, Erving. 1959. *The Presentation of the Self in Everyday Life.* New York: Doubleday.

———. 1974. *Frame Analysis: An Essay on the Organization of Experience.* New York: Harper Colophon.

Gordon, Stewart. 1993. *The Marathas 1600–1818*. Vol.2, part 4 of *The New Cambridge History of India*. Cambridge, UK: Cambridge University Press.

Gould, Harold A. 1990. "Political Economy and the Emergence of a Modern Class System in India." In *Boeings and Bullock-Carts: Studies in Change and Continuity in Indian Civilization*, vol. 1, edited by Yogendra Malik, 154–86. New Delhi: Chanakya Publications.

Gramsci, Antonio. 1971. *Selections from the Prison Notebooks*. Edited and translated by Q. Hoare and G. Nowell Smith. New York: International Publishers.

Guha, Ranajit. 1983. *Elementary Aspects of Peasant Insurgency in Colonial India*. New Delhi: Oxford University Press.

Gupta, Akhil. 1998. *Postcolonial Developments: Agriculture in the Making of Modern India*. Durham, NC: Duke University Press.

Gupta, Dipankar. 1997. *Rivalry and Brotherhood: Politics in the Life of Farmers in Northern India*. New Delhi: Oxford University Press.

Gupta, S. P. 1996. "Recent Economic Reforms in India and their Impact on the Poor and Vulnerable Sections of Society." In *Economic Reforms and Poverty Elimination in India*, edited by C. H. Hanumantha Rao and Hans Linnemann, 126–70. New Delhi: Sage.

Gurr, Ted Robert. 1970. *Why Men Rebel*. Princeton, NJ: Princeton University Press.

Habermas, Jürgen. 1984. *The Theory of Communicative Action*. Boston: Beacon Press.

Hale, Wash Edward. 1999 [1986]. *Asura in Early Vedic Religion*. Delhi: Motilal Banarsidas.

Hall, Edward T. 1969. *The Hidden Dimension*. New York: Doubleday.

Harriss, John. 1982. *Capitalism and Peasant Farming: Agrarian Structure and Ideology in Northern Tamil Nadu*. Bombay: Oxford University Press.

Hart, Gillian. 1991. "Engendering Everyday Resistance: Gender, Patronage and Production Politics in Rural Malaysia." *Journal of Peasant Studies* 19, no. 1 (October): 93–121.

Hartman, Paul; B. R. Patil, and Anita Dighe. 1989. *The Mass Media and Village Life: An Indian Study*. New Delhi: Sage.

Hiltebeitel, Alf, ed. 1989. *Criminal Gods and Demon Devotees: Essays on the Guardians of Popular Hinduism*. Delhi: Manohar.

Hinton, William. 1966. *Fanshen: A Documentary of Revolution in a Chinese Village*. New York: Vintage Books.

Hobsbawm, Eric. 1959. *Primitive Rebels*. New York: W.W. Norton.

Hobsbawm, Eric, and George Rude. 1968. *Captain Swing: A Social History of the Great English Agricultural Uprising of 1830*. New York: W.W. Norton.

Inden, Ronald. 1990. *Imagining India*. Cambridge, MA: Blackwell Publishers.

Irschick, Eugene F. 1994. *Dialogue and History: Constructing South India, 1795–1995*. New Delhi: Oxford University Press.

Jain, Ashok V. 1995. *Government and Politics of Maharashtra*. Bombay: Sheth Publishers.

Jaiswal, Suvira. 1967. *The Origin and Development of Vaisnavism*. Delhi: Munshiram Manoharlal.

Jameson, Fredric. 1981. *The Political Unconscious: Narrative as a Socially Symbolic Act*. Ithaca, NY: Cornell University Press.

Jasper, Daniel. 2002. "Heroic Legacy: Representing and Remembering Shivaji in Maharashtra." PhD diss., New School University, New York, NY.

Jeffery, Patricia, and Roger Jeffery. 1996. *Don't Marry Me to a Plowman! Women's Everyday Lives in North India*. Boulder, CO: Westview Press.

Jhunjhunwala, Ashok. 1983. "Modern Science: A 'Universal Myth.'" In *The Peasant Movement Today*, edited by Sunil Sahasrabudhey, 173–84. New Delhi: Ashish Publishing House.

Joshi, P. C. 1981. "Fieldwork Experience Relived and Reconsidered: The Agrarian Society of Uttar Pradesh." *Journal of Peasant Studies* 8, no. 4 (July): 455–84.

Joshi, Pandit Mahadevashastri. 1962. *Bharatiya Sanskrutikosh* (The Encyclopedia of Indian Culture). Pune: Bharatiya Sanskrutikosh Mandal (in Marathi).

Joshi, Sharad. 1988a. *Shetakari Sanghatana Vicar ani Karyapadhdati* (Thought and Practice of the Shetkari Sanghatana). 3rd ed. Alibaug, India: Shetkari Prakashan (in Marathi).

Joshi, Sharad, Anil Gote, and Rajiv Basergkar. 1988. *Shetakaryanca Raja Shivaji* (Shivaji, King of the Shetkaris). Alibagh, India: Shetkari Prakashan (in Marathi).

Joshi, Tarkateertha Laxmanshastri. 1996. *Jotirao Phule*. New Delhi: National Book Trust.

Karve, Irawati. 1968. *Maharashtra—Land and Its People*. Maharashtra State Gazetteer, General Series. Bombay: Directorate of Government Printing, Stationery and Publications, Maharashtra State.

———. 1988. "'On the Road': A Maharashtrian Pilgrimage." Translated by D. D. Karve and Franklin Southworth. In *The Experience of Hinduism: Essays on Religion in Maharashtra*, edited by Eleanor Zelliot and Maxine

Berntsen, 142–71. Albany: State University of New York Press.

———. 1991 [1961]. *Hindu Society: An Interpretation*. Poona, India: Deccan College, Postgraduate and Research Institution.

Keer, Dhananjay. 1974. *Mahatma Jotirao Phooley: Father of Indian Social Revolution*. Bombay: Popular Prakashan.

Khare, Ravindra S. 1984. *The Untouchable as Himself: Ideology, Identity and Pragmatism among the Lucknow Chamars*. Cambridge, UK: Cambridge University Press.

Kholi, Atul. 1990. *Democracy and Discontent: India's Growing Crisis of Governability*. New York: Cambridge University Press.

Khotari, Rajni. 1986. "Masses, Classes and the State." *Economic and Political Weekly* 21, no 5 (February 1): 210–16.

———. 1990. "The Rise of Peoples' Movements." *Social Action* 40, no. 3 (July–Sept.): 232–40.

King, John Leslie, and Robert L. Frost. 2002. "Managing Distance over Time: The Evolution of Technologies of Dis/Ambiguation." In *Distributed Work*, edited by Pamela Hinds and Sara Kiesler, 3–26. Cambridge, MA: MIT Press.

Kishwar, Madhu. 1992. "Cutting Our Own Lifeline: A Review of Our Farm Policy." *Manushi* 73 (Nov–Dec): 7–25.

———. 2000. "Yes to Sita, No to Ram: The Continuing Hold of Sita on Popular Imagination in India." In *Questioning Ramayanas: A South Asian Tradition*, edited by Paula Richman, 285–308. Berkeley: University of California Press.

Kornhauser, William. 1959. *The Politics of Mass Society*. Glencoe, IL: Free Press.

Kosambi, D. D. 1994 [1962]. *Myth and Reality: Studies in the Formation of Indian Culture*. Bombay: Popular Prakashan.

Kurien, C. T. 1980. "Dynamics of Rural Transformation: A Case Study of Rural Tamil Nadu." *Economic and Political Weekly* 15, no. 5–7, 365–90.

Laclau, E. 1990. *New Reflections on the Revolution of Our Time.* London: Verso.

Laine, James W. 2003. *Shivaji: Hindu King in Islamic India.* New York: Oxford University Press.

Lalini, V. 1991. *Rural Leadership in India.* Delhi: Gian.

Lang, Kurt, and Gladys Lang. 1961. *Collective Dynamic.* New York: Crowell.

Lele, Jayant. 1981. *Elite Pluralism and Class Rule: Political Development in Maharashtra.* Bombay: Popular Prakashan.

———. 1987. "Jnanesvar and Tukaram: An Exercise in Critical Hermaneutics." In *Religion and Society in Maharashtra,* edited by Milton Israel and M. K. Wagle, 115–30. Toronto: Center for South Asian Studies, University of Toronto.

———. 1995. *Hindutva: Emergence of the Right.* Madras: Earthworm Books.

Lele, Jayant, and Rajendra Vora. 1990. "Continuity and Change: Social Basis of Indian Politics." In *Boeings and Bullock-Carts: Studies in Change and Continuity in Indian Civilization,* vol. 5, edited by Jayant Lele and Rajendra Vora, 1–31. New Delhi: Chanakya Publications.

Lenneberg, C. 1988. "Sharad Joshi and the Farmers: The Middle Peasant Lives!" *Pacific Affairs* 61, no. 3, 446–64.

Lévi-Strauss, Claude. 1983 [1969]. *The Raw and the Cooked: Mythologiques,* vol. 1. Translated by John Weightman and Doreen Weightman. Chicago: University of Chicago Press.

Li, Tania Murray. 1996. "Images of Community: Discourse and Strategy in Property Relations." *Development and Change* 27, no. 3 (July): 501–27.

Lincoln, Bruce. 1989. *Discourse and the Construction of Society: Comparative Studies of Myth, Ritual, and Classification.* New Delhi: Oxford University Press.

———. 1999. *Theorizing Myth: Narrative, Ideology, and Scholarship.* Chicago: University of Chicago Press.

Lindberg, Steffan. 1994. "New Farmers' Movements as Structural Response and Collective Identity Formation: The Cases of the Shetkari Sanghatana and the BKU." *Journal of Peasant Studies* 21, no. 3–4 (April–July): 95–125.

———. 1995. "Farmers' Movements and Cultural Nationalism in India: An Ambiguous Relationship." *Theory and Society* 24, 837–68.

Lorenzen, David N. 1987. "The Social Ideologies of Hagiography: Sankara, Tukaram and Kabir." In *Religion and Society in Maharashtra*, edited by Milton Israel and M. K. Wagle, 92–114. Toronto: Center for South Asian Studies, University of Toronto.

Ludden, David. 1992. "India's Development Regime." In *Colonialism and Culture*, edited by Nicholas B. Dirks, 247–88. Ann Arbor: University of Michigan Press.

MacDougall, John. 1980. "Two Models of Power in Contemporary Rural India" *Contributions to Indian Sociology* (n.s.) 14, no. 1, 77–94.

Maharashtra, Government of, Department of Agriculture. 1993. *Agricultural Census 1990–91: Provisional Data on Number and Area of Operational Holdings.* Pune: Government of Maharashtra Department of Agriculture.

Mani, Vettam. 1975. *Purânic Encyclopedia.* Delhi: Motilal Banarsidas.

Mansing, Surjit. 1998. *Historical Dictionary of India*. New Delhi: Vision Books.

Marcus, George E. 1986. "Contemporary Problems of Ethnography in the Modern World System." In *Writing Culture*, edited by James Clifford and George E. Marcus, 165–93. Berkeley: University of California Press.

Marcuse, Herbert. 1960. *Reason and Revolution*. Boston: Beacon Press.

Marglin, Frederique, and Stephen A. Marglin, eds. 1990. *Dominating Knowledge: Development, Culture and Resistance*. Oxford, UK: Clarendon Press

Mathur, H. M., ed. 1977. *Anthropology in the Development Process*. New Delhi: Vikas

Melucci, Alberto. 1985. "The Symbolic Challenge of Contemporary Movements." *Social Research* 52, no. 4. (Winter): 789–815.

———. 1989. *Nomads of the Present: Social Movements and Individual Needs in Contemporary Society*. Philadelphia: Temple University Press.

———. 1996. *Challenging Codes: Collective Action in the Information Age*. Cambridge, UK: Cambridge University Press.

Mencher, Joan. 1974. "The Caste System Upside Down, or the Not So Mysterious East." *Current Anthropology* 15, no. 4, 469–78.

Migdal, Joel S. 1974. *Peasants, Politics and Revolution: Pressures Toward Political and Social Change in the Third World*. Princeton, NJ: Princeton University Press.

Mokashi, D. B. 1987. *Palkhi: An Indian Pilgrimage*. Translated by Philip C. Engblom and Eleanor Zelliot. Albany, NY: State University of New York Press.

Molesworth, James Thomas. 1996 (first published 1831). *Molesworth's Marathi-English Dictionary*. Pune: Shubhada-Saraswat Prakashan.

Moore, Barrington, Jr. 1966. *Social Origins of Dictatorship and Democracy*. Boston: Beacon Press.

Moore, Mick. 1972. "On Not Defining Peasants." *Peasant Studies Newsletter* 1, no. 4 (October): 156–8.

Mudholkar, Aruna and Rajendra Vora. 1984. "Regionalism in Maharashtra." In *Regionalism in India*, edited by Aktar Majeed, 94–97. Delhi: Cosmo.

Nadkarni, M. V. 1987. *Farmers' Movements in India*. Bombay: Allied Publishers.

Nanda, Meera. 2001. "We are All Hybrids Now: The Dangerous Epistemologies of Post-Colonial Populism." *Journal of Peasant Studies* 28, no. 2 (January): 162–186.

Nandy, Ashis. 1987. *Traditions, Tyranny and Utopias: Essays in the Politics of Awareness*. New Delhi, Oxford University Press.

Narayan, Kirin. 1989. *Storytellers, Saints and Scoundrels: Folk Narrative in Hindu Religious Teaching*. Philadelphia: University of Pennsylvania Press.

Nazarea-Sandoval, Virginia D. 1995. *Local Knowledge and Agricultural Decision Making in the Philippines: Class, Gender and Resistance*. Ithaca, NY: Cornell University Press.

Oberoi, Harjat. 1994. *The Construction of Religious Boundaries: Culture, Identity and Diversity in the Sikh Tradition*. Oxford: Oxford University Press.

O'Flaherty, Wendy Doniger. 1976. *The Origins of Evil in Hindu Mythology*. Delhi: Motilal Banarsidass.

———. 1988. *Other Peoples' Myths: The Cave of Echoes.* New York: Macmillan.

O'Hanlon, Rosalind. 1985. *Caste, Conflict, and Ideology: Mahatma Jyotirao Phule and Low Caste Protest in Nineteenth-Century Western India.* Cambridge, UK: Cambridge University Press.

Oliver, Pam, and Hank Johnston. 2000. "What a Good Idea: Frames and Ideologies in Social Movements Research" *Mobilization: An International Journal* 5 (1 April): 37–54.

Omvedt, Gail. 1976. *Cultural Revolt in a Colonial Society: The Non-Brahmin Movement in Western India, 1873–1930.* Bombay: Scientific Socialist Education Trust.

———. 1981. "Capitalist Agriculture and Rural Classes in India." *Economic and Political Weekly* 16, no. 52 (December 26): 140–59.

———. 1993. *Reinventing Revolution: New Social Movements and the Socialist Tradition in India.* Armonk, NY: M. E. Sharpe.

———. 1994a. "Agrarian Transformation, Agrarian Struggles, and Marxist Analyses of the Peasantry." *Bulletin of Concerned Asian Scholars* 26, no. 3 (July–Sept): 47–60.

———. 1994b. "'We Want the Return for Our Sweat': The New Peasant Movement in India and the Formation of a National Agricultural Policy." *Journal of Peasant Studies* 21, no. 3–4 (April–July): 126–64.

Omvedt, Gail, and Bharat Patankar. 2003. "Says Tuka: Songs of a Radical Bhakta" *Critical Asian Studies* 35, no. 2 (June): 277–86.

Ospina, Sonia, and Ellen Schall. 2001. "Perspectives on Leadership: Our Approach to Research and Documentation for the Leadership for a Changing World Program." New York: Leadership for a Changing World.

Padhye, Ramesh. 1985. *Shetakari Andolan* (The Agriculturalists' Struggle). Bombay: Elgar Prakashan (in Marathi).

Paige, Jeffery M. 1975. *Agrarian Revolution: Social Movements and Export Agriculture in the Underdeveloped World*. New York: Free Press.

Panda, Abhash C., and Arun K. Sharma. 1996. "Reconstructing Theory of Social Movements in India" *Social Action* 46, no. 2 (April–June): 111–22.

Pande, Susmita. 1989. *Medieval Bhakti Movement*. Meerut, India: Kusumanjal Prakashan.

Pandey, Gyanendra. 1990. *The Construction of Communalism in Colonial North India*. New Delhi: Oxford University Press.

———. 1992. "In Defense of the Fragment: Writing about Hindu–Muslim Riots in India Today." *Representations* 37 (Winter): 27–55.

Parekh, Bhikhu, and Thomas Pantham, eds. 1987. *Political Discourse: Explorations in Indian and Western Political Thought*. New Delhi: Sage.

Park, Robert E. 1967. *On Social Control and Collective Behavior*. Edited by Ralph H. Turner. Chicago: University of Chicago Press.

Patnaik, Utsa. 1987. *Peasant Class Differentiation: A Study in Method with Reference to Haryana*. New Delhi: Oxford University Press.

Peires, J. B. 1989. *The Dead Will Arise: Nongqawuse and the Great Xhosa Cattle-Killing Movement of 1856–7*. Bloomington: Indiana University Press.

Polletta, Francesca, and James M. Jasper. 2001. "Collective Identity and Social Movements" *Annual Review of Sociology* 27, 283–305.

Popkin, Samuel L. 1979. *The Rational Peasant: The Political Economy of Rural Society in Vietnam*. Berkeley: University of California Press.

Powel, John Duncan. 1972. "On Defining Peasants and Peasant Society." *Peasant Studies Newsletter* 1, no. 3 (July): 94–99.

Prakash, Gyan. 1992. "Writing Post-Orientalist Histories of the Third World: Indian Historiography is Good to Think." In *Colonialism and Culture*, edited by Nicholas B. Dirks, 353–88. Ann Arbor: University of Michigan Press.

Prakash, Gyan, and Douglas Haynes, eds. 1991. *Contesting Power: Resistance and Everyday Social Relations in South Asia*. New Delhi: Oxford University Press.

Putnam, Robert D. 2000. *Bowling Alone: The Collapse and Revival of American Community*. New York: Simon and Schuster.

Race, Jeffrey. 1972. *War Comes to Long An: Revolutionary Conflict in a Vietnamese Province*. Berkeley: University of California Press.

Raheja, Gloria Goodwin, and Ann Grodzins Gold. 1994. *Listen to the Heron's Words: Reimagining Gender and Kinship in North India*. Berkeley: University of California Press.

Ramazanoglu, Caroline. 1993. *Up Against Foucault: Explorations of Some Tensions between Foucault and Feminism*. London: Routledge.

Ranga, N. G. 1988. *The Modern Indian Peasant*. New Delhi: Anmol Publications.

Rao, K. Raghavendra. 1987. "Communication Against Communication: The Gandhian Critique of Modern Civilization in Hind Swaraj." In *Political Discourse: Explorations in Indian and Western Political Thought*, edited by Bhikhu Parekh and Thomas Pantham, 266–76. New Delhi: Sage.

Renan, Ernest. 1990. "What is a Nation?" Translated by Martin Thom. In *Nation and Narration*, edited by Homi K. Bhaba, 8–22. London and New York: Routledge.

Richman, Paula, ed. 1991. *Many Ramayanas: The Diversity of a Narrative Tradition in South Asia*. Berkeley: University of California Press.

———, ed. 2000. *Questioning Ramayanas: A South Asian Tradition*. Berkeley: University of California Press.

Robnett, Belinda. 1997. "African American Women in the Civil Rights Movement: Spontaneity and Emotions in Social Movement Theory." In *No Middle Ground*, edited by Kathleen Blee, 65–95. New York: New York University Press.

Rodrigues, Livi. 1998. *Rural Political Protest in Western India*. New Delhi: Oxford University Press.

Rosaldo, Renato. 1989. *Culture and Truth*. Boston: Beacon Press.

Rudolph, Lloyd I., and Susanne Hoeber Rudolph. 1967. *Modernity of Traditions*. Chicago: University of Chicago Press.

———. 1987. *In Pursuit of Lakshmi: The Political Economy of the Indian State*. Bombay: Orient Longman.

Rushdie, Salman. 1998. "The Mahatma is the Message." *Indian Express*, May 31, 3.

Rutten, Mario. 1995. *Farms and Factories: Social Profile of Large Farmers and Rural Industrialists in West India*. Oxford, UK: Oxford University Press.

Sahasrabudhe, Girish. 1989. "The New Farmers' Movement in Maharashtra." In *Peasant Movement in Modern India*, edited by Sunil Sahasrabudhey, 20–54. Allahabad, India: Chugh Publications.

Sardar, G. B. 1969. *The Saint Poets of Maharashtra: Their Impact on Society*. New Delhi: Orient Longman.

Schlesinger, Lee. 1985. "The Castes in a Village: Caste and a Village." PhD diss., University of Chicago.

———. 1988. "Clan Names, Dominance and Village Organization: Bhosle and Kadam in Apshinge." In *City, Countryside and Society in Maharashtra*, edited by D. W. Attwood, M. Israel and N. K. Wagle, 209–23. Toronto: Center for South Asian Studies.

Scott, James C. 1976. *The Moral Economy of the Peasant: Rebellion and Subsistence in Southeast Asia*. New Haven, CT: Yale University Press.

———. 1985. *Weapons of the Weak: Everyday Forms of Peasant Resistance*. New Haven, CT: Yale University Press.

———. 1990. *Domination and the Arts of Resistance*. New Haven, CT: Yale University Press.

Sharma, Hari P. 1973. "The Green Revolution in India: Prelude to a Red One?" In *Imperialism and Revolution in South Asia*, edited by Kathleen Gough and Hari P. Sharma, 77–102. New York: Monthly Review Press.

Shulman, David Dean. 1980. *Tamil Temple Myths: Sacrifice and Divine Marriage in the South Indian Shaiva Tradition*. Princeton, NJ: Princeton University Press.

Silverman, Sydel. 1983. "The Concept of Peasant and the Concept of Culture." In *Social Anthropology of Peasantry*, edited by Joan P. Mencher, 7–31 Atlantic Highlands, NJ: Humanities Press.

Singh, Prakash. 1995. *The Naxalite Movement in India*. New Delhi: Rupa and Company.

Singh, Yogendra. 1997. *Social Stratification and Change in India*. Delhi: Manohar.

Sirsikar, V. M. 1995. *Politics of Modern Maharashtra*. Bombay: Orient Longman.

———. 1999. "Political Culture of Maharashtra." In *Region, Nationality and Religion*, edited by A. R. Kulkarni and N. K. Wagle, 5–17. Mumbai: Popular Prakashan.

Sivramkrishna, Sashi. 1990. "The Role of Social Movements in Economic Change: Cases from Maharashtra, India." PhD diss., Cornell University.

Skocpol, Theda. 1979. *States and Social Revolutions: A Comparative Analysis of France, Russia and China*. Cambridge, UK: Cambridge University Press.

Smith, Gavin. 1989. *Livelihood and Resistance: Peasants and the Politics of Land in Peru*. Berkeley: University of California Press.

———. 1991. "The Production of Culture in Local Rebellion." In *Golden Ages, Dark Ages: Imagining the Past in Anthropology and History*, edited by Jay O'Brien and William Roseberry, 180–207. Berkeley: University of California Press.

Snow, David A., E. Burke Rochford Jr., Steven K. Worden, and Robert D. Benford. 1986. "Frame Alignment Processes, Micromobilization, and Movement Participation." *American Sociological Review* 51, 464–81.

Snow, David A., and Robert D. Benford. 1988. "Ideology, Frame Resonance, and Participant Mobilization." In *International Social Movement Research: From Structure to Action*, edited by Bert Klandermans, Hanspeter Kriesi, and Sidney Tarrow, 197–218. Greenwich, CT: JAI Press.

Sontheimer, Gunther-Dietz. 1988. "The Religion of the Dhangar Nomads." In *The Experience of Hinduism: Essays on Religion in Maharashtra*, edited by Eleanor Zelliot, and Maxine Berntsen, 109–29. Albany, NY: State University of New York Press.

———. 1997. *King of Hunters, Warriors, and Shepherds: Essays on Khandoba*. Delhi: Manohar.

Spradley, James P. 1979. *The Ethnographic Interview*. Fort Worth, TX: Holt, Rinehart and Winston.

Srinivas, M. N. 1987. *The Dominant Caste and Other Essays*. Oxford, UK: Oxford University Press.

———. 1996. *Village, Caste, Gender and Method: Essays in Indian Social Anthropology*. New Delhi: Oxford University Press.

Stanley, John M. 1987. "Niskama and Sakama Bhakti: Pandharpur and Jejuri." In *Religion and Society in Maharashtra*, edited by Milton Israel, and M. K. Wagle, 51–67. Toronto: Center for South Asian Studies, University of Toronto.

Stree Shakti Sanghatana. 1989. *We Were Making History: Women and the Telangana Uprising*. London: Zed.

Stutley, Margaret. 1993. *Hinduism: The Eternal Law*. New Delhi: Indus.

Suresh, J.K. 1986. "Responses to Technology in the Karnataka Peasant Movement." In *The Peasant Movement Today*, edited by Sunil Sahasrabudhey, 163–72. New Delhi: Ashish Publishing House.

Tarlo, Emma. 1996. *Clothing Matters: Dress and Symbolism in Modern India*. Chicago: University of Chicago Press.

Tarrow, Sidney. 1992. "Mentalities, Political Cultures, and Collective Action Frames." In *Frontiers in Social Movement Theory*, edited by A. D. Morris, and C. M. Mueller, 174–202. New Haven, CT: Yale University Press.

Taussig, Michael. 1980. *The Devil and Commodity Fetishism in South America*. Chapel Hill: University of North Carolina Press.

Thorner, Alice. 1982. "Semi-Feudalism or Capitalism?: Contemporary Debate on Classes and Modes of Production in India" (three parts). *Economic and Political Weekly* 17, no. 49–51 (December 4;11;18): 1961–8; 1993–9; 2061–6.

Tilly, Charles, Louise Tilly, and Richard Tilly. 1975. *The Rebellious Century, 1830–1930*. Cambridge, MA: Harvard University Press.

Touraine, Alain. 1981. *The Voice and the Eye*. Cambridge, UK: Cambridge University Press.

Trouillot, Michel-Rolph. 1991. "Anthropology and the Savage Slot: The Poetics and Politics of Otherness." In *Recapturing Anthropology: Working in the Present*, edited by Richard G. Fox, 17–44. Santa Fe, NM: School of American Research Press.

Turner, Ralph H., and Lewis M. Killian. 1957. *Collective Behavior*. Englewood Cliffs, NJ: Prentice Hall.

Turner, Victor. 1974. "Pilgrimages as Social Processes." In *Dramas, Fields and Metaphors: Symbolic Action in Human Society*, 166–230. Ithaca, NY: Cornell University Press. Originally published as "The Center Out There: Pilgrim's Goal." *History of Religions* 12 (1973): 191–230.

Upadhyaya, Ashok. 1980. "Class Struggle in Rural Maharashtra (India): Towards a New Perspective." *Journal of Peasant Studies* 17, no. 2 (January): 213–34.

Upadhyaya, Carol. 1996. "On Anthropological Discourse." *Economic and Political Weekly* Nov. 30, 3146–48.

Varshney, Ashutosh. 1995. *Democracy, Development and the Countryside: Urban-Rural Struggles in India*. Cambridge, UK: Cambridge University Press.

———. 1997. "Classes, like Ethnic Groups, are Imagined Communities." *Economic and Political Weekly* July 12, 1737–41.

Verma, Manish. 1997. *Fasts and Festivals of India*. New Delhi: Diamond Pocket Books.

Volosinov, V. N. 1986. *Marxism and the Philosophy of Language*. Cambridge, MA: Harvard University Press.

Vora, Rajendra and Suhas Palshikar. 1990. "Neo-Hinduism: A Case of Distorted Consciousness." In *Boeings and Bullock-Carts: Studies in Change and Continuity in Indian Civilization*, vol. 5, edited by Jayant Lele and Rajendra Vora, 213–43. Delhi: Chanakya Publications.

Wallace, Anthony F. C. 1956. "Revitalization Movements." *American Anthropologist* 58, 264–81.

Walton, John. 1987. "Theory and Research on Industrialization." *Annual Review of Sociology* 13, 89–108.

Washbrook, David. 1990. "South Asia, the World System and World Capitalism." *Journal of Asian Studies* 49, no. 3 (Aug.): 479–508.

Weber, Max. 1958. *The Religion of India: The Sociology of Hinduism and Buddhism*. Glencoe, IL: Free Press.

———. 1978 [1968]. *Economy and Society: An Outline of Interpretive Sociology*, vol. 2. Edited by Guenther Roth and Claus Wittich. Berkeley: University of California Press.

Welch, Claude. 1980. *Anatomy of a Rebellion*. Albany, NY: State University of New York Press.

Williams, Raymond. 1973. *The Country and the City*. New York: Oxford University Press.

———. 1976. *Keywords: A Vocabulary of Culture and Society*. n.p: Fontana.

Willis, Paul. 1977. *Learning to Labor: How Working Class Kids Get Working Class Jobs*. New York: Columbia University Press.

Wink, Andre. 1986. *Land and Sovereignty in India: Agrarian Society and Politics under the Eighteenth-Century Maratha Svarajya*. Cambridge, UK: Cambridge University Press.

Wolf, Eric R. 1958. "The Virgin of Guadalupe: A Mexican National Symbol." *Journal of American Folklore* 71, 34–39.

———. 1969. *Peasant Wars of the Twentieth Century*. New York: Harper and Row.

Wolpert, Stanley. 1993. *A New History of India*, 4th ed. New York: Oxford University Press.

Zelliot, Eleanor. 1982. "An Historical View of the Maharashtrian Intellectual and Social Change." In *South Asian Intellectuals and Social Change*, edited by Yogendra Malik, 18–88. New Delhi: Heritage.

———. 1987. "A Historical Introduction to the Warkari Movement." In *Palkhi: An Indian Pilgrimmage*, by D. B. Mokashi, translated by Philip C. Engblom and Eleanor Zelliot, 31–53. Albany, NY: State University of New York Press.

Index

www.ingramcontent.com/pod-product-compliance
Lightning Source LLC
Chambersburg PA
CBHW070557270326
41926CB00013B/2340